Laboratory Manual to accompany

APPLIED PHYSICS

Tenth Edition

Dale Ewen
Parkland Community College

Neill Schurter

P. Erik Gundersen
Pascack Valley Regional High School District

S. Narasinga Rao
University of Central Oklahoma

Prentice Hall

Boston   Columbus   Indianapolis   New York   San Francisco   Upper Saddle River
Amsterdam   Cape Town   Dubai   London   Madrid   Milan   Munich   Paris   Montreal   Toronto
Delhi   Mexico City   Sao Paulo   Sydney   Hong Kong   Seoul   Singapore   Taipei   Tokyo

Editor in Chief: Vernon R. Anthony
Senior Acquisitions Editor: Gary Bauer
Editorial Assistant: Tanika Henderson
Director of Marketing: David Gesell
Marketing Manager: Stacey Martinez
Marketing Assistant: Les Roberts
Senior Managing Editor: JoEllen Gohr

Project Manager: Steve Robb
Senior Operations Supervisor: Pat Tonneman
Art Director: Jayne Conte
Cover Designer: Jeff Vanik
Cover Photo: iStock
Printer/Binder: OPM
Cover Printer: Lehigh-Phoenix Color/Hagerstown

1 2 3 4 V088 13 12 11

Prentice Hall
is an imprint of

ISBN-10:   0-13-210927-1
ISBN-13: 978-0-13-210927-7

PREFACE

In this revised ninth edition of the Laboratory Manual we have included an additional eight experiments on several subjects from mechanics to electricity. These experiments are basic and fundamental to physics and form the basis of a well-balanced laboratory course in physics for a student planning a career in a technical or vocational field. These basics also form a foundation for physics applied in medicine and biology. They are adaptable for use with any basic physics course textbook.

The authors wish to thank Ms. Phyllis Pennington, who has carefully reviewed and prepared the additional experiments for publication.

CONTENTS

Appendices

THE VERNIER CALIPER

Name <u>Virginia Bonivert</u>

Date <u>4/19/11</u>

Period <u> </u>

Lab Partner <u>Amanda F.</u>

Purpose:

The purpose of this lab is to learn to use a vernier caliper. You have studied the use of the English and metric rulers for making measurements. You have found that it is difficult to get very precise results with these instruments. When it is necessary to make measurements which are more precise, you must have a precise instrument. One such instrument is the vernier caliper. It is used by technicians in machine shops, plant assembly lines, and in many other related technical industries. The vernier caliper is a slide-type caliper used to take inside, outside, and depth measurements. The vernier caliper has two metric scales and two English scales (see figure below).

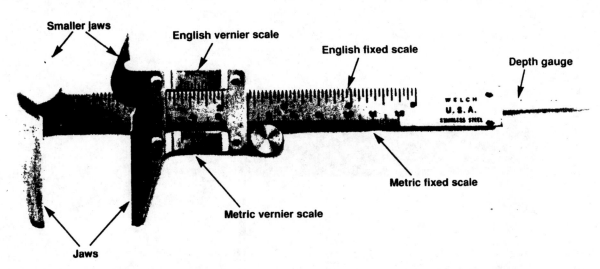

Consider the metric scales. One of them is fixed and located on the lower part of the beam. It is divided into centimeters and subdivided into millimeters, so all readings will be in millimeters. The other metric scale is called the vernier metric scale and is the lower scale on the slide. This scale is made with nine millimeters divided into ten parts. In the next figure, each division on the vernier scale equals 0.9 of a division on the fixed scale. The part of the reading from the vernier scale is in tenths of a millimeter, which means that the precision of the instrument is 0.1 mm or 0.01 cm.

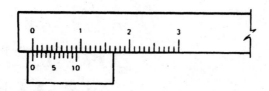

Rules for reading a vernier caliper in metric units:

1. Determine the number of whole millimeters (mm) in the measurement by finding the number of mm graduations on the fixed scale, which are to the left of the first graduation (zero graduation) on the vernier scale. Remember each numbered graduation on the fixed scale represents 10 mm. If the zero graduation is directly in line with a graduation on the fixed scale, the total measurement is read directly from the fixed scale in mm followed by a decimal point and zero.

2. If the zero graduation is not directly in line with a graduation on the fixed scale:

   (a) find the graduation on the vernier scale which is most nearly in line with any graduation on the fixed scale.

   (b) Count the number of graduations on the vernier scale from the zero graduation to determine the number of tenths of millimeters in the measurement.

   (c) Add the numbers from step 1 and step 2(b) to determine the total distance between the jaws of the vernier caliper.

Example 1: Read the measurement on the vernier caliper below in metric units.

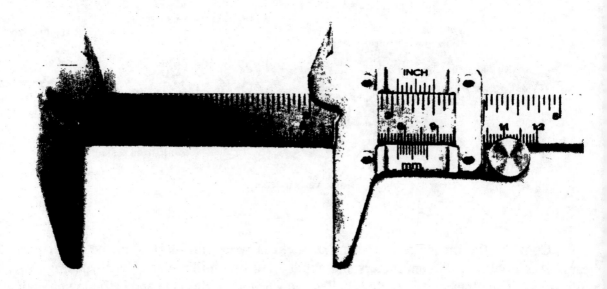

77.0 mm is the first mark to the left of the zero mark.
 0.8 mm is the mark on the vernier that most nearly lines up with a mark on the fixed scale.
77.8 mm is the total measurement.

2

Example 2: Read the measurement on the vernier caliper below in metric units.

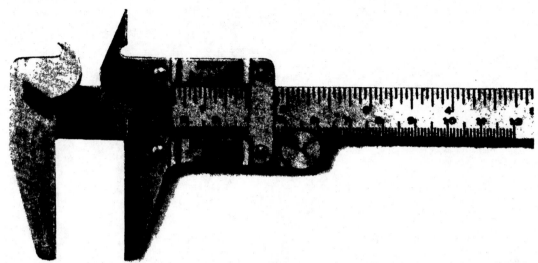

The total measurement is 21.0 mm, since the zero mark on the vernier scale most nearly lines up with a mark on the fixed scale.

Measure the length, width, and height of the rectangular blocks supplied by your instructor. List these dimensions neatly in the table below in mm. Then, using the rules for multiplying measurements, find (a) the surface area of the largest face, and (b) the volume of each block.

Using the platform balance, measure the mass of each block and cylinder. Calculate the density of each. Also calculate the percent difference of densities.

Measure the diameter and height of each cylinder supplied by your instructor. List these dimensions neatly in the table below in mm. Then, using the rules for multiplying measurements, find the volume of each cylinder.

Block	Length	Width	Height	Area of Largest Face	Volume	Mass	Density
1	4.35	2.50	1.4	10.88	15.23	59	0.33
2							
3							
4							
5							

Block	Diameter	Height	Volume	Mass	Density
1					
2					
3					
4					
5					

THE MICROMETER CALIPER

Name _____

Date _____

Period _____

Lab Partner _____

Purpose:

The purpose of this lab is to learn to use a micrometer caliper and to find the gage number of wire.

The micrometer caliper (micrometer or "mike") is a more precise instrument than the vernier caliper and is used in technical fields where fine precision is required. Micrometers are available in metric units and English units. The metric "mike" is graduated and is read in hundredths of a millimeter (0.01 mm). The English "mike" is graduated and is read in thousandths of an inch (0.001 in.).

The basic parts of a micrometer are labeled in the figure below. The object to be measured is placed between the anvil and spindle. Turn the thimble until the object fits snugly. <u>Do not force the turning of the thimble, since this may damage the very delicate threads on the spindle located inside the thimble.</u> Some calipers have a ratchet, which helps protect the instrument by not allowing the thimble to turn when forced.

<u>Part 1</u>: **Metric micrometer**

The following figure shows a metric micrometer in which the basic parts are labeled.

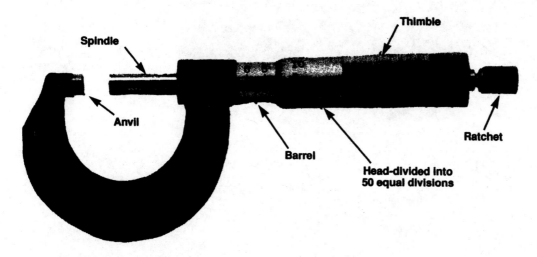

The barrels of most metric micrometers are graduated in millimeters. The micrometer above also has graduations in halves of millimeters, which are indicated by the lower set of graduations on the barrel. The threads on the spindle are made so it takes two complete turns of the thimble for the spindle to move precisely one millimeter. The head is divided into fifty equal divisions—each division indicating 0.01 mm, which is the precision of the instrument.

Rules for reading a metric micrometer in millimeters:

1. Find the <u>whole number</u> of mm in the measurement by counting the number of mm graduations on the barrel to the left of the head.

2. Find the <u>decimal part</u> of the measurement by reading the graduation on the head that is most nearly in line with the center line on the barrel, and multiply this reading by 0.01. If the head is at or immediately to the right of the half mm graduation, add 0.50 mm to the reading on the head.

3. Add the numbers found in step 1 and step 2.

Example 1: Read the measurement on the metric micrometer below.

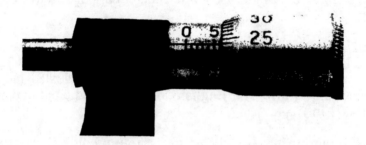

1.	The barrel reading is	6.00 mm
2.	The head reading is	<u>0.24 mm</u>
3.	The total measurement is	6.24 mm

Example 2: Read the measurement on the metric micrometer below.

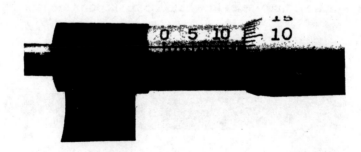

1.	The barrel reading is	14.00 mm
2.	The head reading is	<u>0.08 mm</u>
3.	The total measurement is	14.08 mm

Example 3: Read the measurement on the metric micrometer below.

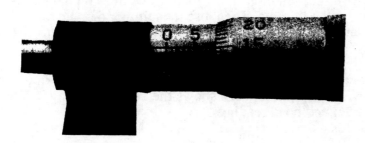

1. The barrel reading is   8.00 mm
2. The head reading is   <u>0.65 mm</u>   (Note that the head is past
3. The total measurement is   8.65 mm   the half mm mark.)

Measure the diameter of the nails supplied by your instructor. List the diameters (in mm) neatly below.

Diameter (mm)

#1

#2

#3

#4

#5

Part 2: Metric micrometer with vernier scale

By adding a vernier scale on the barrel of the micrometers we just studied, we can increase the precision by one more decimal place, that is, the metric micrometer with vernier scale has precision 0.001 mm, while the English version has precision 0.0001 inch. We will discuss the metric version here. The following figure shows a metric micrometer with vernier scale.

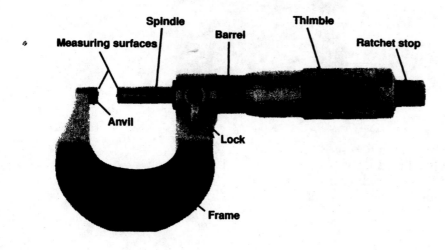

7

This micrometer is read like the metric micrometer at the beginning of this experiment, except that an additional reading in thousandths of a millimeter from the vernier scale is added. To get the vernier reading, find which of the lines on the vernier scale most nearly line up with a line on the thimble. Then, use the number marked on that vernier line as the number of thousandths of a millimeter, which is then added to the hundredths reading.

Example 1: Read the measurement on the metric micrometer with vernier scale below. Note: In the figure below, the top photograph shows the vernier reading, while the bottom photograph shows the barrel and head readings.

1. The barrel reading is	6.000 mm
2. The head reading is	0.350 mm
3. The vernier reading is	<u>0.006</u> mm
4. The total measurement is	6.356 mm

Example 2: Read the measurement on the metric micrometer with vernier scale below.

1. The barrel reading is	8.500 mm
2. The head reading is	0.150 mm
3. The vernier reading is	<u>0.007</u> mm
4. The total measurement is	8.657 mm

Measure the diameter of the nails supplied by your instructor. List the diameters (in mm) neatly in the following table.

Diameter (mm)

1

2

3

4

5

Part 3: **Gage number of copper wire**

Electricians choose the sizes of electrical wires according to the amount of current a given circuit will use. The size of the wire used must be increased as the amount of current in the circuit increases. All wires are made into standard sizes according to their diameters. These standard sizes are classified by gage numbers. Electricians use gage numbers of wire rather than the diameters to classify the sizes of wires.

Table 19 in the Appendix 9 lists the sizes of wires according to gage numbers and diameters in both metric units and English units. The metric units are in millimeters, and the English units are in mils (1 mil = 0.001 inch).

Measure and record the diameters of the copper wires supplied by your instructor. Then find the gage number of each wire from Table 19. (Do not force the thimble, as this not only may damage the mike but also may flatten the wire.)

	Diameter	Gage Number		Diameter	Gage Number
1.			7.		
2.			8.		
3.			9.		
4.			10.		
5.			11.		
6.			12.		

<u>Part 4</u>: **Conversion factors**

You may be faced with the problem of having only an English micrometer when you need to make a precise measurement on a Japanese motorcycle or other vehicle with measurements in the metric system.

1 mm = 0.0394 in. = 39.4 X 10^{-3} in. = 39.4 thousandths in.

1 in. = 25.4 mm (exactly)

Example: Convert 0.355 in. to millimeters and round the result to three significant digits.

Using a conversion factor, we have

$$0.355 \text{ in.} \times \frac{25.4 \text{ mm}}{1 \text{ in.}} = 9.02 \text{ mm}$$

Questions:

Convert each measurement to millimeters and round the result to three significant digits.

1. 0.025 in. _____   2. 0.037 in. _____

3. 0.049 in. _____   4. 0.104 in. _____

5. 0.308 in. _____

Convert each measurement to inches and round the result to three significant digits.

6. 6.05 mm _____   7. 0.609 mm _____

8. 3.90 mm _____   9. 0.406 mm _____

10 . 8.20 mm _____

10

ACCELERATION

Experiment

3

Name_____

Date_____

Period_____

Lab Partner_____

(handwritten) g = accel. gravity $9.8 \frac{m}{s^2}$
a = accel. object

Purpose:

The purpose of this lab is to measure the acceleration of a ball rolling down an inclined track.

Equipment   Inclined track
                  Steel ball
                  Meterstick or yardstick
                  Timer

(diagram of inclined track with $9.8 \frac{m}{s^2}$ labeled)

$\theta = \sin^{-1}\dfrac{(a)}{(g)}$

$a = g \sin \theta$

Note: $a = g \sin \theta$ holds good for a ball rolling down a smooth (frictionless) inclined track. For a rolling solid ball, the formula is $a = 0.71\, g \sin \theta$

Procedure:

1. Release the ball from the top of the inclined track and mark its position at one second intervals. Several trials will be necessary.

2. Record in the table the distance between your marks, which is the distance traveled in each second.

3. Calculate and record the average velocity during each second using $v = s/t$.

4. Calculate and record the velocity change between seconds.

5. Acceleration is the velocity change divided by the time required for the change. Calculate and record the acceleration between seconds.

6. Draw a graph of position versus time.

7. Draw a graph of velocity versus time.

8. Draw a graph of acceleration versus time.

9. Calculate the slope in each case.

10. Calculate the angle of the track.

(three graph axes: Position vs Time, Velocity vs Time, Acceleration vs Time)

Time Intervals	Distance Traveled		v, Velocity, $\Delta S/\Delta t$	Velocity Change, Δv	Acceleration $\Delta v/\Delta t$
	S	ΔS			
1st second	.08 m	.08 m	.08 m/s		
2nd second	.29 m	.21 m	.21 m/s	.13 m/s	.13 m/s²
3rd second	.72 m	.43 m	.22 m/s	.21 m/s	.22 m/s²
4th second	.96 m	.24 m	.08 m/s	.16 m/s	m/s²
5th second	m	m	m/s	m/s	m/s²
6th second	m	m	m/s	m/s	m/s²

$$Vf - Vi =$$

$$\frac{.13}{2-1} = .13 \frac{m}{s^2}$$

Questions:

1. How should the accelerations be related?

    They are all increasing.

2. List as many factors as you can that might affect the accuracy of this experiment.

    How high or low the incline is, speed of the ball.

3. Suggest two ways you think this experiment could be improved.

    Having a specific incline, surface use, high speed camera (freeze frame).

4. What is the difference between "distance" and "displacement"?

    Distance, displacement - one is scalar and one is vector.

5. Why should the unit of time appear twice in the unit for acceleration?

6. Define and distinguish between the terms "acceleration" and "deceleration" (retardation). Acceleration- speeding up.

    deceleration- slowing down.

7. What is the difference between the terms "speed" and "velocity"?

    speed = displacement per unit time
    velocity - speed w/a direction vector.

8. What is the difference between a "scalar" and a "vector" quantity?

    Scalar=
    vector=

9. Identify "distance" and "displacement; "speed" and "velocity" as a scalar or a vector quantity.

10. How would you define speed? How would you define velocity?

THE PENDULUM

Experiment **4**

Name _____

Date _____

Period _____

Lab Partner _____

Purpose:

The purpose of this lab is to find whether or not the material a pendulum is made of affects its rate of vibration (period), to find how the length of a pendulum affects its rate of vibration, and to find whether or not the length of the arc affects the rate of vibration.

Equipment:

Wooden ball
Metal ball
String
Meterstick
Stopwatch or watch with second hand
Pendulum support

<u>Note</u>: This experiment may be done either in pairs of students or as a class demonstration.

Procedure:

A simple pendulum consists of a small mass suspended by a string about a point which allows the pendulum to vibrate freely. See the figure below. We will call the point about which the pendulum vibrates point P. As the pendulum swings from point A to point B and back again to point A, we say that it completes one complete vibration or one complete cycle. The time required for one complete cycle is called the period. The number of cycles per second is called its frequency. The displacement of a pendulum is its distance from point C. Arc AC represents the maximum displacement. The length of the pendulum is distance PC.

1. Set up the pendulum as shown by attaching a string to a ball and to a clamp on some support.

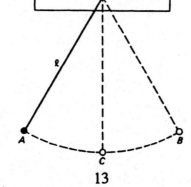

2. Fasten a wooden ball to a piece of string so that the length of the pendulum, PC, is 1.00 meter long. Pull the ball so that the string makes about 10^0 with the vertical. Then release it to allow it to begin swinging. One complete oscillation is the swing from A through C to B and back to A through C. When the ball is released at A, start the stopwatch. Note the time for 10 complete oscillations. Calculate the period, or the time required for one oscillation. Complete the table for Trial 1.

3. Repeat step 2, using a metal ball. Complete the table for Trial 2.

4. Using a metal ball, calculate the time for 10 oscillations and hence, the period for a pendulum 1.00 m, 75 cm, 65 cm, 50 cm, 35 cm, 25 cm, and 10 cm long. Complete the table for Trials 3-9.

5. Using a metal ball, make the pendulum 1.00 in. long. Displace the pendulum 15 cm, 20 cm, and 30 cm and calculate the time for 10 oscillations. Complete the table for Trials 10-12.

6. Using the results in Trials 3-9, plot the graph of period versus length. Also draw the graph of time versus length on a log-log graph.

7. Plot the graph of T^2 versus L. The graph of L versus T^2 will be a straight line with slope $g/4\pi^2$ from which g can be calculated.

Trial	Material	Amplitude (m)	$L=\dfrac{T^2g}{4\pi^2}$ (m)	Length (m)	$T=2\pi\sqrt{\dfrac{L}{g}}$ (s)	Time for 10 Complete Oscillations	Period (s)
1	Wood	0.10		1.00			
2	Metal	0.10		1.00			
3	Metal	0.10		1.00			
4	Metal	0.10		0.75			
5	Metal	0.10		0.65			
6	Metal	0.10		0.50			
7	Metal	0.10		0.35			
8	Metal	0.10		0.25			
9	Metal	0.10		0.10			
10	Metal	0.15		1.00			
11	Metal	0.20		1.00			
12	Metal	0.30		1.00			

Questions:

1. How is the period of the pendulum affected by the kind of material used?

2. How is the period of the pendulum affected by its length?

3. How is the period of the pendulum affected by the amplitude of vibration?

4. Interpret the result of graph T versus L on the linear graph.

5. Interpret the graph of T versus L on the log-log graph.

6. Calculate the slope of the graph in question 5.

7. Calculate the percentage error between the calculated and the theoretical value of the slopes.

8. Calculate the acceleration due to gravity from T^2 versus L.

9. In 1602, Sanctorius, a medical friend of Galileo, invented the pulsologium (a simple pendulum) to measure the pulse rate of his patients. Using this idea of Sanctorius, determine the "length" of your pulse rate.

FRICTION Experiment

5

Name _____

Date _____

Period _____

Lab Partner _____

Purpose:

The purpose of this lab is to study the force that resists the sliding of one object over another—friction. We will determine the coefficient of friction for some surfaces and determine the effects of changing the weight and surface area of our materials.

Equipment:

Block of wood with hook attached
Spring scale
200-g and 500-g masses
Platform balance

Procedure:

1. Determine whether your spring scale actually reads zero when no mass is attached. If not, you will have to correct each reading by subtracting the amount of error from each reading.

2. Find the mass of the wood block and record it in the table that follows.

3. Determine the weight of the block using $F_w = mg$.

4. Place the block on a flat, horizontal surface with the largest surface area down.

5. Pull the block at uniform speed along the surface with the scale and read the scale while still pulling the block. Do this after the reading steadies but while the block is still moving. Be careful to keep the scale parallel to the surface.

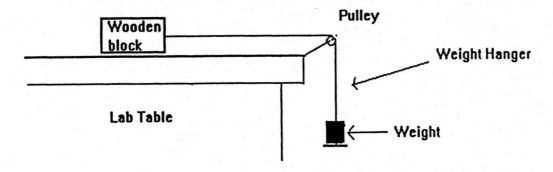

6. Repeat this procedure at least three times until consistent readings are obtained. Then find the force of friction by converting the mass scale reading using $F = mg$ and record it in the table.

7. Now calculate the coefficient of friction using

$$\mu = \frac{F_f}{F_N}$$

where μ (Greek lowercase letter mu) = the coefficient of friction

$\qquad F_f$ = the frictional force

$\qquad F_N$ = the normal force pressing the objects together (here, the weight of the block)

Enter your result in the table, rounded to two decimal places.

8. Repeat the above procedure using the same block with at least two different masses added but with the block in the same position so the area between surfaces remains the same. Record your results.

9. Repeat the above procedure in each case but with the block turned on its side so you are using a different area of contact between the surfaces. Record your results.

10. Finally, if your instructor so directs, repeat the above procedure on a different horizontal surface.

Trial Number	Zero Reading of Spring Scale (1)	Reading of Scale with the Object (2)	Mass of the Object (2)-(1)	$F_N = mg$	Scale Reading while Pulling F_f	$\mu = \frac{F_f}{F_N}$
1	0	750	750g	7.35N	.5N	.068
2						
3						
4						
5						
6						

Questions:

1. Does the coefficient of sliding friction increase, decrease, or remain the same as the load increases?

    Decrease

2. Does μ increase, decrease, or remain the same if the area of contact between the surfaces is reduced? It remains the same because we didn't reduce the friction but were consistant.

3. Can you suggest a reason why the scale reading is larger as the block begins to move as compared to later, when it is uniformly sliding?
    Because of the force applied.

4. List at least three ways friction between sliding surfaces can be reduced.
    By polishing surfaces, converting sliding friction to rolling friction, using liquid and powderd lubricants.

5. Distinguish between the terms: "Precision" and "Accuracy".
    Precision: of a measurment systems is the degree to which repeated under unchanged conditions show same results.
    Accuracy: of a measurement system is the degree of closeness of measurments of a quantity to its actual true value.

6. From the calculated values of μ in the last column, can you predict your results as being precise or accurate or both?
    Precise measurments remaind the same and showed the same results each time.

7. Suppose you do not record the zero reading of the spring scale. What would happen to your results? What does this tell you about "Calibration" of the instrument before you use it?
    If we didn't record the zero reading or didn't check for calibration our results would be different and we would probably have to subtract the difference.

19

TRIGONOMETRY

Experiment

6

Name _Virginia Bonivert_

Date _4/26/11_

Period _____

Lab Partner _____

Purpose:

The purpose of this lab is to become familiar with the use of basic trigonometric ratios to indirectly measure distance.

Equipment:

    Protractor
    Tape measure
    Centimeter scale

Procedure:

Find the trigonometric ratios for the 48° angle in triangle A and triangle B using:

$$\sin 48° = \frac{\text{side opposite to } 48°}{\text{hypotenuse}}$$

$$\cos 48° = \frac{\text{side adjacent to } 48°}{\text{hypotenuse}}$$

$$\tan 48° = \frac{\text{side opposite to } 48°}{\text{side adjacent to } 48°}$$

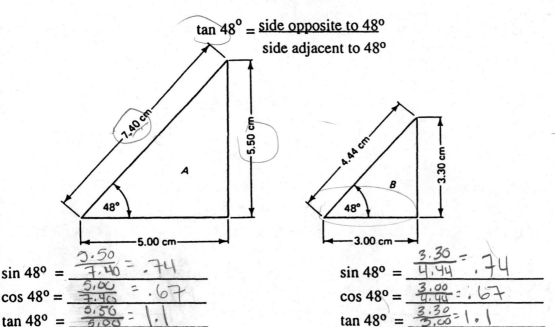

$\sin 48° = \dfrac{5.50}{7.40} = .74$

$\cos 48° = \dfrac{5.00}{7.40} = .67$

$\tan 48° = \dfrac{5.50}{5.00} = 1.1$

$\sin 48° = \dfrac{3.30}{4.44} = .74$

$\cos 48° = \dfrac{3.00}{4.44} = .67$

$\tan 48° = \dfrac{3.30}{3.00} = 1.1$

Round the values of the ratios to these decimal places.

<u>Note</u>: The trigonometric ratios of 48° are the same in both triangles.

Measure the legs and hypotenuse of triangles A and B below with a centimeter scale. Then find the trigonometric ratios of the 30° angle using:

$$\sin 30° = \frac{\text{side opposite to } 30°}{\text{hypotenuse}}$$

$$\cos 30° = \frac{\text{side adjacent to } 30°}{\text{hypotenuse}}$$

$$\tan 30° = \frac{\text{side opposite to } 30°}{\text{side adjacent to } 30°}$$

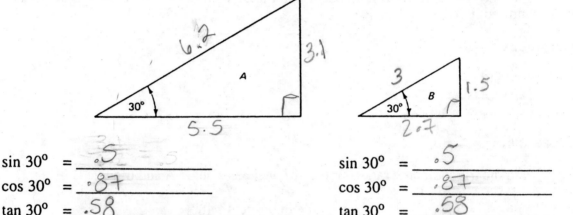

sin 30° = .5   = .5   sin 30° = .5
cos 30° = .87   cos 30° = .87
tan 30° = .58   tan 30° = .58

Round the values of the ratios to these decimal places.

Note: The trigonometric ratios of 30° are the same in both triangles.

Draw a right triangle using a protractor and a centimeter scale so that one angle is 37° and its adjacent side is 9 cm. Then draw another right triangle so that one angle is 37° and its adjacent side is 6.5 cm. Then find the trigonometric ratios of the 37° angle of each triangle.

$$\sin 37° = \frac{\text{side opposite to } 37°}{\text{hypotenuse}}$$

$$\cos 37° = \frac{\text{side adjacent to } 37°}{\text{hypotenuse}}$$

$$\tan 37° = \frac{\text{side opposite to } 37°}{\text{side adjacent to } 37°}$$

sin 37° = .60   sin 37° = .60
cos 37° = .80   cos 37° = .80
tan 37° = .75   tan 37° = .75

Note: The trigonometric ratios should be the same in both right triangles if you were careful in drawing the triangles. The trigonometric ratios of a specific angle are always the same in any right triangle.

22

Questions:

1. A surveying party wishes to find the distance between a point (*A*) on one side of a stream and a point (*B*) on the opposite Bank. The surveyor sets the transit over point *A* and backsights on point *B*. He then turns his transit through a 90° angle and foresights toward point *C*. The party sets point *C* at any convenient distance from *A* and, using a steel tape, measures and records a distance of 801 ft. The surveyor sets the transit over point *C*, backsights on *A*, and turns the telescope to point *B*. He measures the angle between his sighting on *A* and his sighting on *B* and finds it is 28°. With this information, what is the distance from point *A* to point *B*?

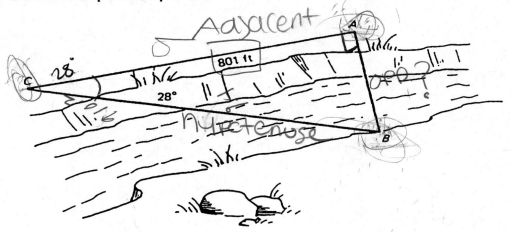

2. Find the distance between two trees or other objects using a method like the one used in the river problem above. (Use a protractor and a tape measure.)

3. List below all of the sources of error you can think of that might affect the accuracy of this experiment.

4. Why must you be sure you have a right angle?

5. List any other situations you can think of where an indirect measurement like the one you made here might be essential or useful.

6. Calculate the distance *AB* and *BC*, using trigonometric formulas.

7. Measure *AB* and *BC*. Compare the results to the above.

EQUILIBRIUM OF CONCURRENT FORCES

Experiment 7

Name _____

Date _____

Period _____

Lab Partner _____

Purpose:

The purpose of this lab is to study an object (an iron ring) in equilibrium under the action of several concurrent forces. We will find the x and y components of each force vector and show that the sum of the x components = 0 and the sum of the y components = 0.

Equipment:

Two spring balances
Iron ring (about 1/2" in diameter)
Cord
Weight hangers with slotted weights
Rod and clamps

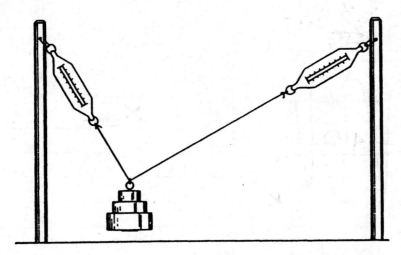

Procedure:

1. Attach three pieces of cord to the iron ring and to the spring balances and weight hanger as shown.

2. Add enough weights to the hanger so that the scale readings are a least one-third of full scale. Record the scale readings and the hanging mass. (Remember to use $F_w = mg$ to convert to weight.)

1.12 lbs

3. Using a protractor, measure and record the angle of each force in standard position.

4. Draw the free-body diagram in the space provided.

5. Find the **x and y components** of each force using the component method.

6. Find the sum of the x components and the sum of the y components. These two sums should be very close to zero if you were careful.

7. Change the length of cord between one balance and the iron ring to change the bearing of the forces. Repeat steps 2-6.

Free-Body Diagrams:

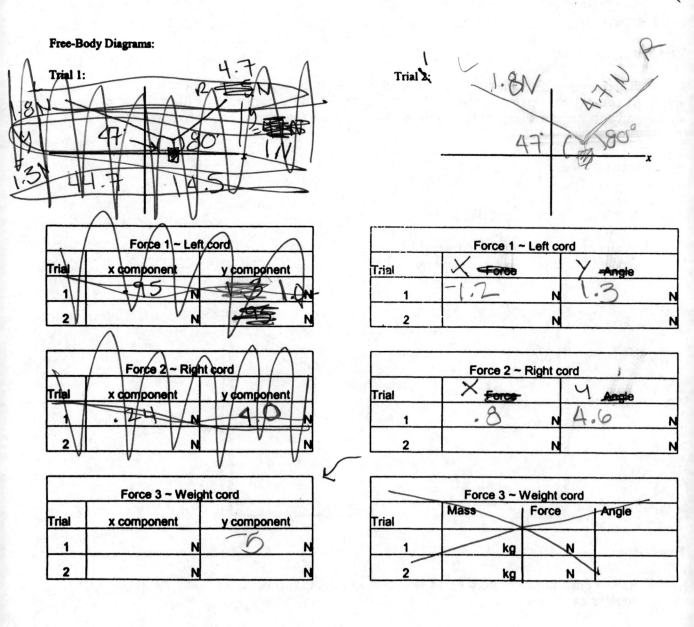

Trial 1:

Trial 2:

Force 1 ~ Left cord		
Trial	x component	y component
1	.95 N	1.0 N
2	N	N

Force 2 ~ Right cord		
Trial	x component	y component
1	.24 N	4.0 N
2	N	N

Force 3 ~ Weight cord		
Trial	x component	y component
1	N	5 N
2	N	N

Force 1 ~ Left cord		
Trial	X ~~Force~~	Y ~~Angle~~
1	-1.2 N	1.3 N
2	N	N

Force 2 ~ Right cord		
Trial	X ~~Force~~	Y ~~Angle~~
1	.8 N	4.6 N
2	N	N

Force 3 ~ Weight cord		
Trial	Mass	Force	Angle
1	kg	N	
2	kg	N	

26

Trial	Total Components	
	sum of x components	sum of y components
1	−0.4 N	o9 N
2	N	N

Questions:

1. Have we verified the equilibrium conditions within a reasonable margin of error? State the conditions of equilibrium.

2. List the sources of error in this experiment.

SOH

T

EQUILIBRIUM OF THE CRANE BOOM

Name _____

Date _____

Period _____

Lab Partner _____

Purpose:

The purpose of this lab is to study the equilibrium conditions involved in supporting objects with a crane.

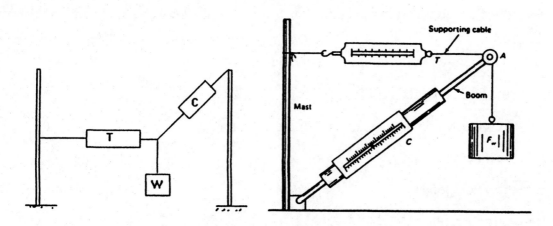

We are interested in three forces: the weight supported, F_w; the compression in the boom, C; and the tension in the supporting cable, T. Since all these forces (concurrent forces) act on the point labeled A, we isolate this point and draw the forces acting on it in a free-body diagram. Since A is in equilibrium (no acceleration), the sum of the x components and the sum of the y components of all the forces should be zero.

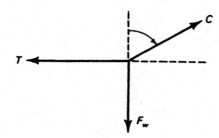

Equipment:

Crane boom with internal spring balance for measuring compression
Spring balance for measuring tension in supporting cable
Weights
Protractor
Cord

Procedure:

Determine whether your spring scale actually reads zero when no mass is attached. If not, you will have to correct each reading by subtracting the amount of error from each reading.

1. Set up the boom, supporting cable, and spring balance as shown.

2. Add the weight, F_w. Convert mass to weight using $F_w = mg$ and record.

3. Adjust the supporting cable so that it is horizontal.

4. Record the compression in the boom. Measure the angle between the boom and the mass.

5. Record the tension in the supporting cable and its bearing.

6. Draw the free-body diagram for each trial.

7. Find the x and y components of each force vector.

8. Add the x components.

9. Add the y components.

10. Repeat for one other weight.

11. Adjust the supporting cable so that it is not horizontal (make a positive angle with the horizontal). For the same load, record tension and compression force.

12. Repeat step 11 by making a negative angle with the horizontal. For the same load, record tension and compression force.

Free-body Diagrams:

Trial 1:   Trial 2:   Trial 3:

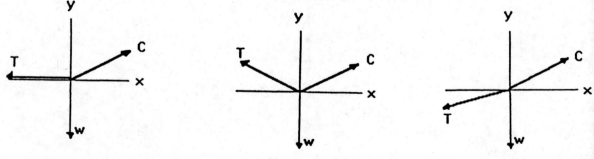

Data:

Force F_w			
Trial	Mass	Force	Angle
1	kg	N	
2	kg	N	
3	kg	N	

Force F_w		
Trial	x component	y component
1	N	N
2	N	N
3	N	N

Force C			
Trial	Mass	Force	Angle
1	kg	N	
2	kg	N	
3	kg	N	

Force C		
Trial	x component	y component
1	N	N
2	N	N
3	N	N

Force T			
Trial	Mass	Force	Angle
1	kg	N	
2	kg	N	
3	kg	N	

Force T		
Trial	x component	y component
1	N	N
2	N	N
3	N	N

Total Components		
Trial	sum of x components	sum of y components
1	N	N
2	N	N
3	N	N

Questions:

1. Have we verified the equilibrium conditions in this experiment?
   (Sum of x components = 0 and sum of y components = 0.)

2. How do you think the x components and y components change if the supporting cable is shortened so the boom is more nearly upright?

3. List at least five sources of error.

HORSEPOWER/ MANPOWER

Experiment

9

Name _____

Date _____

Period _____

Lab Partner _____

Purpose:

The purpose of this lab is to measure the horsepower of an electric motor and/or the "manpower" of a student climbing a flight of stairs.

Equipment:

Electric motor (1/20 to 1/4 hp) with pulley
Revolutions counter
Two spring scales
Heavy cord or belt
Stopwatch
Hand rotator
Flight of stairs

Part 1: Manpower

Time a student running up a flight of stairs. Repeat, having other students walk up or run at different speeds. The distance measured will be the height of the stairs. What is the force ?

Compute the horsepower developed using the relationships:

Power $\quad = \dfrac{\text{Work}}{\text{Time}}$ (in ft lb/s)

Work $\quad = \text{Force} \times \text{Distance}$

Horsepower $\quad = \dfrac{\text{Power}}{550 \frac{\text{ft/lb}}{\text{s}}}$

	Height of Stairs	Weight of Runner	Time to Ascend	Hp Developed
Student #1	ft	lb	s	hp
Student #2	ft	lb	s	hp
Student #3	ft	lb	s	hp

Part 2: Horsepower-Electric Motor

Set up the apparatus as shown below.

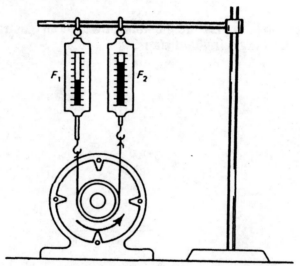

The net force on the wheel is the difference between the scale readings. The distance traveled is the circumference of the pulley times the number of revolutions in the timed interval.

$$\text{Work} = 2\pi r \times (F_1 - F_2) \times N$$

where

r = radius of pulley
F_1, F_2 = scale readings
N = number of revolutions

then

$$\text{Horsepower} = \frac{Work}{Time \times 550 \, \dfrac{ft/lb}{s}}$$

Student	Pulley Radius	F_1	F_2	Time	No. of Revolutions	Horsepower
=1	ft	lb	lb	s		hp
=2	ft	lb	lb	s		hp
=3	ft	lb	lb	s		hp

Part 3: **Horsepower-Hand Rotator**

Set up the apparatus as in Part 2, substituting the hand rotator for the electric motor. Adjust the belt so you can turn the rotator rapidly but with some effort for 30 seconds.

Questions:

1. Why did we use the vertical height of the stairs for the distance?

2. List at least two different ways to measure the height of stairs.

3. How is it possible that a student might have the fastest time up the stairs and still not develop the most horsepower?

35

pg. 269

$$MA = \frac{\text{effort arm}}{\text{resistance arm}} = \frac{Se}{Sr}$$

(mean)
AdV

$$= \frac{3m}{1m}$$

$$\boxed{= 3}$$

$F_R S_R = F_e S_e$ $Fe = 1$ $2 = F_R?$

$2 \cdot 1 = 1 \cdot 2$

$2 = 2$

       $\downarrow$

       $\underline{\hspace{2cm}}$ $\triangle$ $SR = 1$

       $Se = 2$

$Fe = \text{your effort } F'$

$Se = \text{your effort lever arm}$

      SUd

$F_R = \text{resistence } F$

$S_R = \text{resistance lever arm}$

$\downarrow \quad\quad \downarrow$

$\underline{\hspace{3cm}}$ "seesaw"

   $\triangle$ 1st class lever

$\uparrow\downarrow$

$\underline{\hspace{2cm}}\big|$ "wheel barrow"

     $\triangle$ 2nd class lever

$\uparrow$

$\underline{\hspace{2cm}}$ "tweezer"

$\downarrow$ $\triangle$ 3rd class lever

LEVERS

Experiment

10

Name _____

Date _____

Period _____

Lab Partner _____

Purpose:

The purpose of this lab is to learn about the three classes of levers.

Equipment:

Meterstick
Spring scale
Weights
String

Procedure:

Set up examples of the three types of levers.

First Class

Second Class

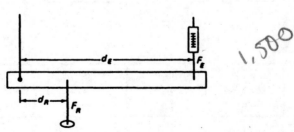

1,500

Third Class

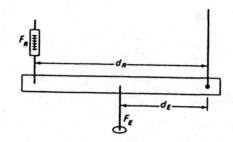

Vary the distance and resistance and redo the experiment four times for each class. Also, check the actual mechanical advantage using the forces with the calculated MA (using the distances).

First Class

Trial	Effort Force F_E	Effort Distance d_E	Resistance Force F_R	Resistance Distance d_R	AMA (Actual)	CMA (Calculated)	Effort: % AMA/CMA × 100
1	20.0	11cm	500	4cm			
2							
3	2N	11cm	5N	4cm	2.5	2.75	90.90
4							

Second Class

F=ma

Trial	Effort Force F_E	Effort Distance d_E	Resistance Force F_R	Resistance Distance d_R	AMA (Actual)	CMA (Calculated)	Effort: % AMA/CMA × 100
1	3N	15cm	2N	24cm			
2							
3	3N	15cm	2N	24cm	0.67	0.63	104.35
4							

Third Class 209

Trial	Effort Force F_E	Effort Distance d_E	Resistance Force F_R	Resistance Distance d_R	AMA (Actual)	CMA (Calculated)	Effort: % AMA/CMA × 100
1	2N	23.5cm	2.8N	20 cm	14	1.18	1186.44
2							
3							
4							

Questions:

1. Does the position of the fulcrum affect the mechanical advantage of the lever?

2. Diagram each of the three classes of levers in the space below.

   First Class

   Second Class

   Third Class

3. How does the actual MA compare to the calculated MA (using the distances)?

4. Which of the levers multiplies force?

5. Which of the levers multiplies distance or speed?

SIMPLE MACHINES

Name _____

Date _____

Period _____

Lab Partner _____

Purpose:

The purpose of this lab is to study the wheel and axle, pulleys, the inclined plane, and the screw.

Part 1: **Wheel and axle**

Equipment:

> Wheel and axle apparatus (may have to be made)
> Spring scale
> Weights
> String

Procedure:

Set up a wheel and axle situation as shown in **Figure 1.** Make three trials using various weights. Then alter the apparatus as shown in **Figure 2.** Make three more trials. Compare the calculated MA (using the radius of the wheel and the radius of the axle) with the actual MA (using the resistance and the effort forces) for each trial.

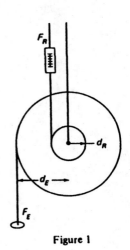

Figure 1

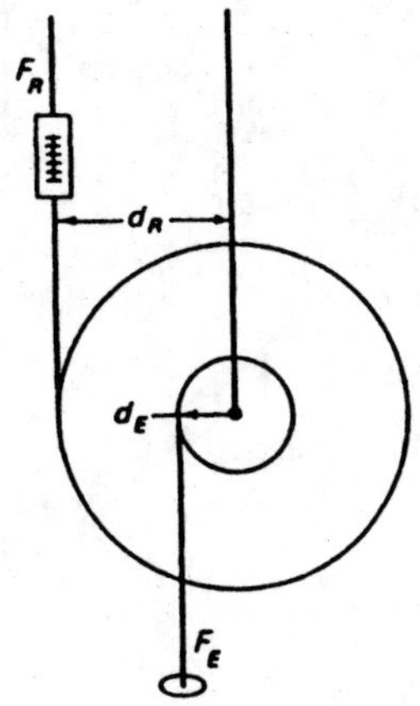

Figure 2

Trial	Resistance Force F_R	Effort Force F_E	Resistance Radius r_R	Effort Radius r_E	AMA (Actual)	CMA (Calculated)	Effort: % AMA/CMA× 100
1							
2							
3							
4							
5							
6							

Which MA is larger calculated or actual? Why?

Part 2: **Pulleys**

Equipment:

Pulleys
Spring scale
Weights
String

Procedure:

Set up some pulley systems, as shown in the following figures, or with available equipment, if different pulley systems are available. Measure the effort and resistance forces. Measure the distances the resistance and effort forces move in relation to each other.

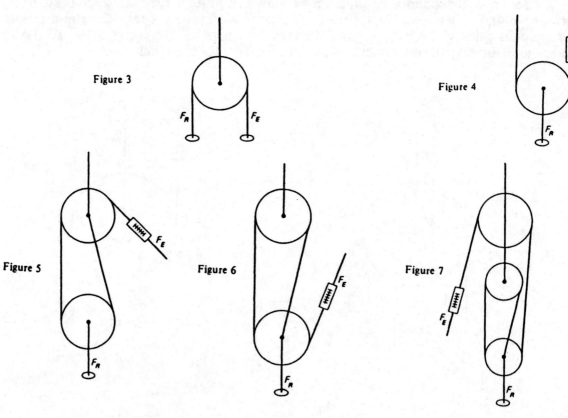

Figure 3

Figure 4

Figure 5

Figure 6

Figure 7

Compare the calculated MA (number of supporting strands) with the actual MA (using the resistance and effort forces). Calculate also the ratio of the resistance distance to the effort distance.

Trial	Resistance Force F_R	Effort Force F_E	Resistance Distance d_R	Effort Distance d_E	CMA (Calculated)	AMA (Actual)	Effort: % AMA/CMA× 100
1							
2							
3							
4							
5							

Part 3: Inclined Plane

Equipment:

Board and pulley
Car
Weights
String

Procedure:

Set up several trials of the inclined plane by varying the length and height of the plane and the resistance.

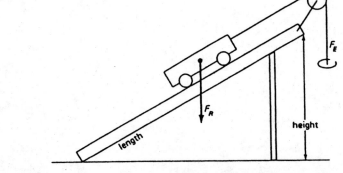

Measure the resistance force, effort force, resistance distance, and effort distance. Also, compare the actual and calculated MA. Find the actual MA using the effort and resistance forces. Find the calculated MA using the effort and resistance distances.

Trial	Length of Plane	Height of Plane	Resistance Force F_R	Effort Force F_E	CMA (Calculated)	AMA (Actual)	Effort: % AMA/CMA× 100
1							
2							
3							
4							
5							

Part 4: Screw

Equipment:

Jackscrew
Board to sit on
Bathroom scale or other appropriate scale

Procedure:

Set up an experiment using a jackscrew to lift some heavy object (student). Vary the resistance and length of the handle where possible to vary the experiment. Determine the resistance force, effort force, resistance distance, and effort distance. Also, compare the calculated MA (using the pitch and length of the handle) with the actual MA (using the resistance force and effort force).

Trial	Resistance Force F_R	Effort Force F_E	Pitch	Radius	AMA (Actual)	CMA (Calculated)	Effort: % AMA/CMA× 100
1							
2							
3							
4							
5							

CENTRIPETAL FORCE

Name _____

Date _____

Period _____

Lab Partner _____

Purpose:

The purpose of this lab is to study the nature of centripetal force and the relationship of the force, mass, and velocity.

$$F = \frac{mv^2}{r}$$

Equipment:

Glass tube about 15 cm long
Fishline
Rubber stopper
Set of masses
Tape
Stopwatch or watch with
    second hand

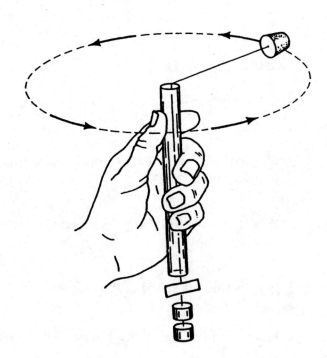

In this experiment we will change each factor in the equation as the stopper is whirled.

Procedure:

Trial 1. Begin by whirling a rubber stopper (approximately 50-g mass), maintaining a velocity so that the piece of tape moves neither up nor down. (See figure.) After some practice in holding the piece of tape steady, find the period (time) of one rotation by timing 40 revolutions and dividing the total time by 40. Record your data and complete the table on the next page for Trial 1, M, hanging mass = 50 gm.

Trial 2. Follow the same procedure using a 70-g hanging mass this time.

Trial 3. Follow the same procedure using a 100-g hanging mass this time.

Trial 4. Follow the same procedure using a different stopper.

Trial 5. Using the same stopper, vary the radius.

Recall:

<u>Weight</u>   <u>Circumference of Circle</u>

$$F_w = mg \qquad\qquad C = 2\pi r$$

$$F_{centripetal} = \frac{(\text{Mass of stopper})\,(\text{Velocity})^2}{\text{Radius}}$$

Data:

Mass of stopper 1 _____

Mass of stopper 2 _____

Questions:

1. Was the centripetal force of the hanging weight close to the centripetal force found by calculation?

2. Describe what you think the motion of the stopper would be if the string were cut.

3. How would the speed of the stopper change if you increase the hanging mass?

4. How did the speed of the stopper change when you increased the mass of the stopper?

Trial	Mass on String M	Centripetal Force (Weight of mass on string) $F_w = 9.8\,M$	Radius r	Circumference C $(2\pi r)$	Time for 40 rev	Period t	Velocity (C/t) v $r\omega$ C/t	Velocity v^2	Centripetal Force (Calculated) $\dfrac{M\,stopper\ v^2}{r}$	ω rad/s
1	0.050 kg	N	m	m	s	s	m/s	(m/s)²	N	
2	0.070 kg	N	m	m	s	s	m/s	(m/s)²	N	
3	0.100 kg	N	m	m	s	s	m/s	(m/s)²	N	
4	0.050 kg	N	m	m	s	s	m/s	(m/s)²	N	
5	0.050 kg	N	m	m	s	s	m/s	(m/s)²	N	

PARALLEL FORCES AND EQUILIBRIUM

Experiment **13**

Name Virginia Bonivert

Date _____

Period _____

Lab Partner_____

Purpose:

The purpose of this lab is to illustrate some problem situations you faced in solving torque problems.

In the lab we will show the two conditions for equilibrium:

1. The vector sum of the forces must be zero (sum of forces up equals sum of forces down). ($\Sigma Fy = 0$)

2. The sum of the clockwise torques equals the sum of the counterclockwise torques. ($\Sigma M_O = 0$)

Equipment:

    Meterstick
    Platform balance
    Two spring balances
    Weights
    String

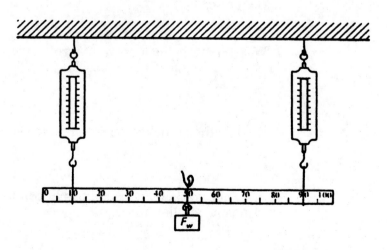

In this experiment we will vary the size and location of the hanging weight and verify the two conditions of equilibrium. Record all your data in the table.

Since your equipment includes spring balances, which are probably calibrated in grams, and a set of masses, it will be necessary to change grams to kilograms and multiply by the acceleration of gravity (g = 9.80 m/s^2) to determine the weight of the masses used and the forces. (Recall that $F_w = mg$.)

Procedure:

Part 1:   *500g weight*

1. Make sure the spring scale reads zero in the vertical position before suspending the meterstick. If not note the reading on the scale and a proper correction needs to be applied.

2. Determine the weight of the meterstick. (This weight is present in all trials as a downward force at its center of gravity- the 50-cm mark.)

3. With the weight at the 50-cm mark, add various weights (your instructor will tell you which ones) and read both spring scales. Record your data in **Table 1** on the next page. Do both scales read the same? Do you think they should?

4. Move the weight to the 30-cm mark and repeat step 3.

5. Move the weight to the 75-cm mark and repeat step 3.

6. Calculate the percentage difference between total weight and sum of the scale forces.

Note:

The center of gravity of the meterstick is at the 50 cm mark only if the meterstick is uniform. Otherwise the center of gravity shifts. We assume that the meterstick is uniform and the center of gravity is at the 50 cm mark.

Questions:

1. In a sentence or two, compare the total weight and the sum of the two forces shown on the scales. *The sum of the 2 forces adds up to almost equal the total weight.*

2. With a reasonable amount of error, have we shown that the sum of the forces up equals the sum of the forces down?
   Yes.

3. Explain the term "moment of a force about a point."
   you torque of the pivot.

4. State and explain the conditions of equilibrium of a body acted on by a number of coplanar parallel forces.
   Sum of all forces is O

+5.2 N
-5.17

$\dfrac{.03}{5.17}$

TABLE 1

Trial	Mass of Meter Stick 27.69	Weight of Meter Stick kg 27.69	Weight of Meter Stick N	Weight of Mass Added kg	Weight of Mass Added N	Total Weight	Sum of Forces Shown on Scales Left kg	Left N	Right kg	Right N	Total N	% Diff.
1	.0276 kg	.0276	.27048 N	.5	5 N	5.1704 N		2.5 N		2.7 N	.03	0.0058
2	kg		N		N	N		N		N		
3	kg		N		N	N		N		N		
1	.0276 kg	.0276	.2704 N	.5	5 N	5.1704 N		3.5 N		1.7 N	.03	"
2	kg		N		N	N		N		N		
3	kg		N		N	N		N		N		
1	kg		N		N	N		1.3 N		3.9 N	.03	"
2	kg		N		N	N		N		N		
3	kg		N		N	N		N		N		

At 50-cm mark (trials 1–3)
At 30-cm mark (trials 1–3)
At 75-cm mark (trials 1–3)

The second condition for equilibrium is that $T_{clockwise} = T_{counterclockwise}$. For this part we will use the weight (force) data obtained in Part 1.

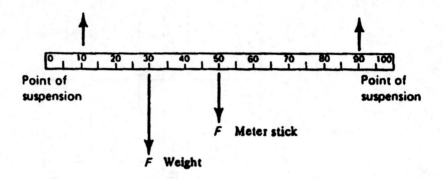

Procedure:

1. Use the left-hand point of suspension (while facing the equipment) as the point of rotation. Measure all distances from this point.

2. Calculate each torque present when the weight is at the indicated positions, and complete Table 2 on the next page.

Questions:

1. Are the clockwise and counterclockwise torques nearly equal? Should they be?

   Yes, nearly equal.

2. When the meterstick is balanced, how are the forces related to the lengths of the arms on which they act? The longer the lever there is less force, The shorter, The greater The force.

3. Are the torques affected if some other point is chosen for the axis of rotation? If so, how?
   Where you put the pivot changes everything.

4. Would the data be affected if the spring balances were attached upside down?
   Data would not be affected.

TABLE 2

	Clockwise Torques ($F \cdot s_l$)			Counterclockwise Torques ($F \cdot s_l$)	
Trial	Meterstick	Hanging Weights	Total Clockwise Torque	Force at Right-Hand Support	Total Counterclockwise Torque
At 50-cm mark 1	8.8 N m	1.96 N m	.56 N m	1.4 N	.56 N m
2	N m	N m	(1.4N x .4m) N m	N	(1.4 x .4m) N m
3	N m	N m	N m	N	N m
At 30-cm mark 1	8.3 N m	N m	.38 N m	.9 N	.54 N m
2	N m	N m	(1.9N x .2m) N m	N	(.9N x .6m) N m
3	N m	N m	N m	N	N m
At 75-cm mark 1	.8 N m	N m	.46 N m	2 N	.3 N m
2	N m	N m	(.7N x .65m) N m	N	(2N x .15m) N m
3	N m	N m	N m	N	N m

HOOKE'S LAW

Experiment **14**

Friction

Name _____

Date _____

Period _____

Lab Partner _____

Purpose:

The purpose of this lab is to show that distortion of an elastic body is proportional to the distorting force if the elastic limit is not exceeded. (The distance stretched depends on the amount of weight, unless you stretch it so far that it won't return to its original shape.)

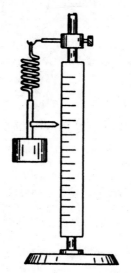

Equipment:

Hooke's Law apparatus or miscellaneous bars, clamps, and springs
Meterstick
Set of masses

We will attempt to verify Hooke's Law: $\dfrac{F}{\Delta l} = k$

where: F = weight on spring
   Δl = distance spring is stretched
   k = experimentally determined elastic constant of the spring

Procedure:

1. Adjust the scale or hanger so that the pointer reads zero on the scale.

2. Hang at least six (preferably ten) masses on the hanger and record the distance the spring is stretched (elongation) for each mass. Your instructor will suggest which masses to use for your particular springs.

3. Repeat using a different spring.

4. Draw a graph of elongation along the X axis versus force (weight) along the Y axis.

5. Draw the best straight line through the data points.

6. Calculate the slope of the line that gives the elastic constant of the spring.

Questions:

1. Do you think Hooke's Law might also apply to bending? If so, think of and describe below an experiment to verify your answer.

   Bending a spring, it is only supposed to elongate and stretch not bend.

2. Why are spring balances unreliable for precise work?

   Because springs lose elacticity.

3. How would the results of this experiment be different if it were conducted in space? On the moon? *There is no gravity.*

4. What can you say about the relationship between force and elongation from your graph? *you increase length you increase force.*

5. What is the effect of force, extension, and elastic constant in space and on the moon? *The gravitational force is different.*

6. Calculate the percentage difference between the slopes obtained from graphs and the elastic constants obtained from calculations.

	Trial	Mass		Mass		Weight		Elongation		Elongation		ELASTIC CONSTANT	
Spring 1	1	40	g	.040	kg	0.39	N	4.9	cm	0.049	m	7.96	N/m
	2		g		kg		N		cm		m		N/m
	3		g		kg		N		cm		m		N/m
	4		g		kg		N		cm		m		N/m
Spring 2	1	50	g	.050	kg	.49	N	4.4	cm	0.064	m	7.66	N/m
	2		g		kg		N		cm		m		N/m
	3	20	g	.02	kg	.196 N	N	2.6	cm	.026	m	7.31	N/m
	4	30	g	.030	kg	.294	N		cm	.632	m	7.06	N/m

*Convert mass given in grams to kilograms to related weight (newtons).

$k = \frac{F(N)}{\Delta l(m)}$

$T = k$

$k = $ spring stretch

spring constant

force exerted

Use a suitable scale and graph your data.

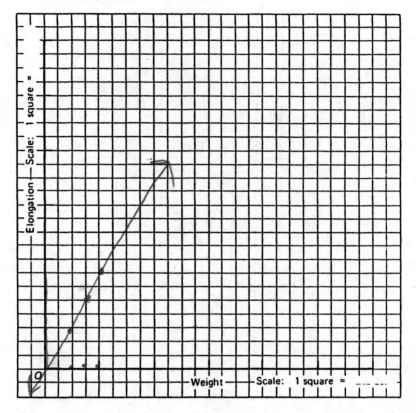

Connect the springs as shown. Then find the relationship for k_T in terns of k_1 and k_2 in each case.

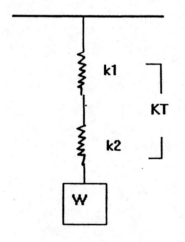

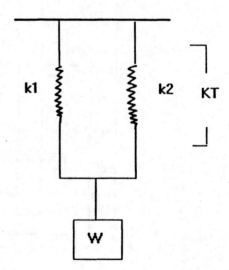

ARCHIMEDES' PRINCIPLE

Experiment

15

Name _____

Date _____

Period _____

Lab Partner _____

Purpose:

The purpose of this lab is to show that the weight of the liquid displaced by a submerged body is equal to the buoyant force that the liquid exerts on that body.

Equipment:

Overflow can
Catch bucket
Platform balance
Large beaker or battery jar
Metal block
Thread

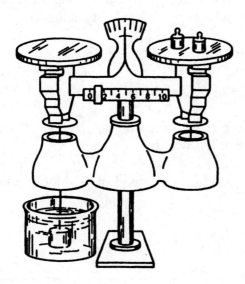

Procedure:

Objects that are less dense than water float in water due to the upward force exerted by the water. Those objects that are more dense than water seem to lose some of their weight when submerged in water. This is also due to an upward force exerted by the water. We will measure this upward force (buoyant force) and compare it to the weight of the displaced water.

1. Attach the metal block to the platform balance with a light thread.

2. Carefully find the mass of the block and calculate its weight. Record this value in the table.

3. Lower the block into the beaker of water so that it is completely below the surface and does not touch the sides or bottom of the beaker.

4. Record the apparent mass and weight of the block while submerged.

5. Find the upward force (buoyant force) exerted on the block by subtracting its weight in water from its weight in air.

6. To find the weight of the water displaced by the block, fill the overflow can to the spout. Set the empty catch can below the spout. Holding the thread, slowly lower the block into the overflow can until it is completely submerged.

7. Find and record the weight of the catch bucket and water.

8. Find and record the weight of the catch bucket alone.

9. Subtract the weight of the bucket from the combined weight of the bucket and water to fmd the weight of the displaced water.

10. Repeat procedure steps 1-9 with the second metal block of equal volume but different density.

11. Calculate the density of the block.

Data:

Trial	Mass of Block in Air	Weight of Block in Air	Apparent Mass of Block in Water	Apparent Weight of Block in Water
1	g	N	g	N
2	g	N	g	N

Apparent Loss in Weight (Buoyant Force)	Weight of Displaced Water and Catch Bucket	Weight of Catch Bucket	Weight of Displaced Water gm		% Difference
N	N	N	N	N	
N	N	N	N	N	

Questions:

1. Is the buoyant force equal to the weight of the displaced water?

2. Why does the block weigh less when immersed in water?

3. Compare the weight of the block to the weight of an equal volume of water.

4. Why is the buoyant force the same for both metal blocks?

5. Calculate the density of each metal block with the help of the data you have recorded in the table. (Density = mass of metal block in air divided by the apparent loss of mass in water.) Compare these values to the accepted values for the metal. Calculate the percentage error.

6. Calculate the percentage difference between buoyant force and the weight of displaced water.

7. Why does the water exert a buoyant force on the metal block?

SPECIFIC GRAVITY OF SOLIDS

Experiment **16**

Name _____

Date _____

Period _____

Lab Partner _____

Purpose:

The purpose of this lab is to determine the specific gravity of a solid whose weight density is greater than water and of a solid whose density is less than water.

Equipment:

Platform balance
Battery jar
Sinker
Thread
Pieces of rock, metal, cork, etc.

Specific gravity is the ratio of the density of an object to the density of water. Since the block and displaced water have the same volume, we may find specific gravity by comparing the mass of the block to the mass of the water.

Procedure:

To find the specific gravity of a solid that sinks in water, use the formula:

$$\text{Specific gravity of a solid} = \frac{\text{Mass of solid in air}}{\text{Mass of water displaced}}$$

Solids More Dense Than Water (Those That Sink):

1. Use a solid whose density is greater than that of water.

2. Find the mass of the solid in air using a platform balance.

3. Tie a thread to the solid and submerge it in water. Do not let it touch the bottom or sides of the battery jar. Catch the overflow.

4. Find the mass of the water that overflowed.

5. Calculate the specific gravity of the solid by using the formula above.

6. Make two trials for each material used.

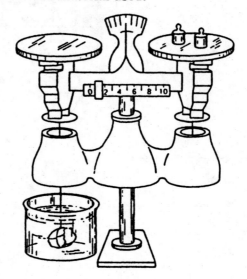

Data:

Material	Trial	Mass of Block in Air	Apparent Mass of Submerged Block	Mass of Water Overflow	Specific Gravity Calculated Chart	% Difference
	1	g	g	g		
	2	g	g	g		
	1	g	g	g		
	2	g	g	g		
	1	g	g	g		
	2	g	g	g		

Solids Less Dense Than Water (Those That Float):

1. Use a solid whose weight density is less than water.

2. Weigh the solid in air using a platform balance.

3. Tie a string to the solid and a sinker as shown on the next page.

4. Using a platform balance, find the mass of the solid out of the water with the sinker under water. Make sure the sinker is not touching the bottom or sides of the jar and the solid is above the water. Discard the water overflow.

5. Lower <u>both</u> sinker and block below the water and <u>catch the overflow.</u>

6. Find the mass of water overflow. This is the mass of water that has the same volume as the block.

7. Calculate the specific gravity as before by comparing the mass of the block to the mass of the water.

8. Make two trials for each material used.

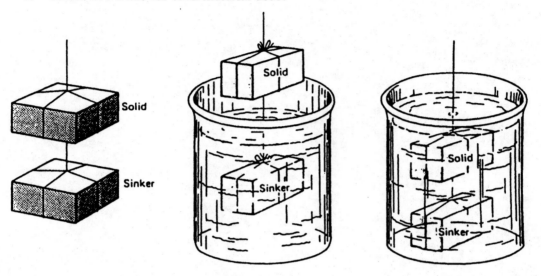

Data:

Material	Trial	Mass of Block in Air	Mass of Water Overflow	Specific Gravity
	1	g	g	
	2	g	g	
	1	g	g	
	2	g	g	
	1	g	g	
	2	g	g	

Questions:

1. Why was it necessary to tie a sinker to the floating solid?

2. List three possible sources of error in your work.

3. Explain the difference between "Specific Gravity" and "density".

4. The density of water is 1 gm/cm^3. What is its density in the British system?

SPECIFIC GRAVITY OF LIQUIDS

Name _____

Date _____

Period _____

Lab Partner _____

Purpose:

The purpose of this lab is to determine the specific gravity of a liquid. We will use two different methods.

Equipment:

Platform balance
Several liquids (milk, carbon tetrachloride, alcohol, oil, gasoline, etc.)
Battery jar, glass stopper
Hydrometers, hydrometer jars
Thread and water

Procedure:

Method 1. Loss-of-weight method:

We may find specific gravity (ratio of densities) by comparing the buoyant effects of liquids on a submerged object.

$$\text{Specific gravity} = \frac{\text{Apparent loss of mass in liquid}}{\text{Apparent loss of mass in water}} = \frac{\text{Wt in air - Wt in liquid}}{\text{Wt in air - Wt in water}}$$

1. Pour a liquid into a battery jar.

2. Tie a glass stopper on a piece of thread.

3. Find the mass of the glass stopper in air using a platform balance. Weigh glass stopper in air.

4. Find the mass of the stopper in water. Weigh glass stopper in water.

5. Find the mass of the stopper in the liquid in which you are finding the specific gravity.

6. Find the apparent loss of mass of the stopper in the water and then in the liquid.

7. Find the specific gravity of the liquid.

67

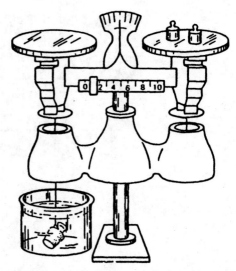

Liquid Used	Trial	Mass of Stopper			Apparent Loss of Mass		Specific Gravity
		in Air	in Water	in Liquid	in Water	in Liquid	
	1	g	g	g	g	g	
	2	g	g	g	g	g	
	1	g	g	g	g	g	
	2	g	g	g	g	g	
	1	g	g	g	g	g	
	2	g	g	g	g	g	

Method 2. Hydrometer method:

1. Fill a hydrometer jar with a liquid so that a hydrometer will not touch the bottom.

2. Find the depth (reading) of the hydrometer.

3. Carefully wipe all liquid off the hydrometer before placing it in another liquid.

Liquid Used	Specific Gravity

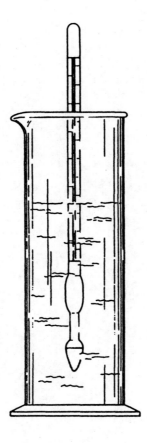

Questions:

1. Do your specific gravities found by the hydrometer method closely match those found by the loss-of-weight method?

2. What principle is involved in the use of the hydrometer?

3. What is the order of the numbers on the hydrometer scale? Why are they placed that way?

SPECIFIC HEAT

Experiment 18

Name _____

Date _____

Period _____

Lab Partner _____

Purpose:

The purpose of this lab is to determine the specific heats of several different metals.

Equipment:

Styrofoam cup (calorimeter)
Celsius thermometer
Beaker
Burner and rubber tubing
Set of masses
Platform balance
Various metal blocks
String

Procedure:

Each metal block is heated to the boiling point of water (about 100°C) and then placed into a calorimeter containing water. The heat lost by the metal block is equal to the heat gained by the water and the calorimeter. Since the calorimeter is composed of styrofoam, the heat it gains is so small we can neglect it. We then have:

Heat lost by metal block = Heat gained by water

$$c_l m_l (T_l - T_f) = c_g m_g (T_f - T_g)$$

1. Begin by boiling water in a beaker half full of water.

2. Find the mass of the metal block and record it in the table.

3. Attach a piece of string about 15 inches long to the block and lower it into the beaker.

4. Find and record the mass of the empty calorimeter cup.

5. Fill the cup about one-half full of water that is a few degrees colder than room temperature, and find the mass of the cup and water. Subtract the mass of the cup to obtain the mass of the water in the cup. Best results will be obtained if the approximate temperature of the water is as far below room temperature when starting as it is above room temperature after the metal block has been added.

6. Find and record the temperature of the boiling water. All temperature measurements should be to the nearest tenth of a degree.

7. Find and record the temperature of the water in the cup.

8. Holding the solid by the string, support the metal in the steam just above the water. Leave it there until the water on the block has evaporated.

9. Quickly lower the block into the calorimeter.

10. Stir the water with the stirring rod.

11. Record the temperature of the water when it reaches its highest point.

12. Perform the calculations necessary to fill in the data below. You will use the equation given above for some calculations.

Data:	Trial 1	Trial 2
Kind of metal	_____	_____
m_1, Mass of metal (2)	_____ g	_____ g
Mass of calorimeter (4)	_____ g	_____ g
Mass of calorimeter and water (5)	_____ g	_____ g
m_g, Mass of water (5)	_____ g	_____ g
T_2, Initial temperature of metal (T_1) (7)	_____ °C	_____ °C
T_g, Initial temperature of water in the cup (8)	_____ °C	_____ °C
T_f, Final temperature of metal and water (12)	_____ °C	_____ °C
Temperature change of water ($T_f - T_g$)	_____ °C	_____ °C
Q_g, Calories (heat) gained by water	_____ cal	_____ cal
Total calories (heat) lost by metal, $Q_2 = Q_g$	_____ cal	_____ cal
Temperature change of metal, ($T_2 - T_f$)	_____ °C	_____ °C
C_2, Specific heat of metal (calculated)	_____ cal/g°C	_____ cal/g°C
C, Accepted value for specific heat (from Table 15 on page 201)	_____ cal/g°C	_____ cal/g°C
Error	_____ cal/g°C	_____ cal/g°C

$$Q_g = C_g m_g (T_f - T_g)$$

$$\% \text{ error} = \frac{C_2 - C}{C} \times 100$$

$$C_2 = \frac{Q_g}{m_2 (T_2 - T_f)}$$

Questions:

1. Why should the water initially be several degrees cooler than room temperature?

2. What are the sources of error in this experiment?

3. How does the specific heat of water compare to the various metals?

4. Why should the metal block be held out of the boiling water until most of the water clinging to it falls off?

Note: Explanation concerning item 5(on page 72). We were trying to minimize the radiation effects to and from the surrounding ambient temperature.

STANDING WAVES

Name _____

Date _____

Period _____

Lab Partner _____

Purpose:

The purpose of this lab is to study standing waves on a vibrating string. We will find the conditions necessary to obtain several different standing wave patterns for a given vibration frequency.

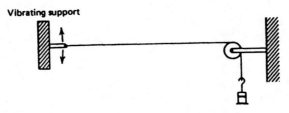

Equipment:

  Vibrating support
  String
  Weight hanger with slotted weights
  Pulley
  Rods and clamps
  Meterstick
  Balance

Theory:

    If a uniform string is subjected to a tension, and one end of the string is given a vibratory motion at right angles to the length of the string, then waves travel along the length of the string. The motion of any particle of the string is at right angles to the length of the string. The waves travel along the length of the string. The direction of motion of any particle of the string is at right angles or transverse to the direction in which the waves travel. Such waves are called "transverse" waves. The waves consist of a regular succession of "crests" and "troughs" traveling down the string. The highest point of the string above the undisturbed line is called "crest" and the lowest point is called "trough." The <u>amplitude</u> of the waves is the highest point of a crest or the depth of a trough measured from the equilibrium point. The <u>frequency</u> of the waves is the number of crests or troughs that pass any point on the string per second. The <u>period</u> is the time interval between two successive crests or two successive troughs. The frequency f and the period T are related by

$$f = \frac{1}{T}$$

The distance between two successive crests or two successive troughs is one <u>wavelength</u> λ. The velocity of the wave V is given by

$$V = \frac{\lambda}{T}$$

The velocity of the transverse wave is also given by

$$V = \sqrt{\frac{T}{\mu}}$$

where T is the tension (force) in the string and μ is the mass per unit length of the string.

Therefore:

$$V = \sqrt{\frac{T}{(m/l)}}$$

Squaring:

$$V^2 = \frac{Tl}{m}$$

$$\lambda^2 f^2 = \frac{Tl}{m}$$

$$T = f^2 \frac{m}{l} \lambda^2$$

This equation can be used to determine the frequency of the wave on a string.

Procedure:

1. Attach a length of string over the pulley, with one end of the string connected to the vibrating support and the other end connected to the weight hanger as shown on page 79.

2. Plug the vibrator into an electrical outlet and switch on.

3. Begin adding weights to the hanger until a stable standing wave pattern is obtained.

4. Record the required weight.

5. Measure and record the wavelength of the standing wave pattern.

6. Determine the wave velocity from $V = \lambda f; V = \lambda \times 120$.

7. The velocity of the wave on the string is given by $V = \sqrt{T/\mu}$, where T is the tension in the string and μ is the mass per unit length of the string. Using this equation and your experimental data, calculate the mass per unit length of the string.

8. Repeat the above procedure at least two more times by adding more weight and producing different standing wave patterns.

9. Calculate the average value for μ from your data.

10. Measure the length and weight of a similar piece of string. Determine μ from these measurements.

11. Plot a graph of the tension versus the square of the wavelength. Draw the best straight line through the points. The slope of this line is $f^2(m/l)$. Determine the slope. Use the equation to calculate the frequency of the vibrating support.

12. Find the percentage error between the frequency you can determine and correct frequency of the vibrator.

Trial	T	λ	v	μ
1	N	m	m/s	kg/m
2	N	m	m/s	kg/m
3	N	m	m/s	kg/m

Questions:

1. Does the value of μ determined directly from the length and mass measurement in step 9 agree with the average value determined in step 8?

2. What are the possible sources of error in this experiment?

SPEED OF SOUND

Experiment

21

Name _____

Date _____

Period _____

Lab Partner _____

Purpose:

The purpose of this lab is to determine the speed of sound in air.

Equipment:

Tuning fork
Air column with adjustable height
Meterstick

Procedure:

1. Adjust the water level to a position close to the top of the glass tube.

2. Strike the tuning fork and hold it over the column as shown below.

3. Lower the water level slowly until a resonance (maximum sound level) is observed. Record the height of the air column.

4. The first resonance corresponds to one quarter of a wavelength as shown. From this fact, the equation $v = \lambda f$ and the known frequency of the tuning fork determine the velocity of sound.

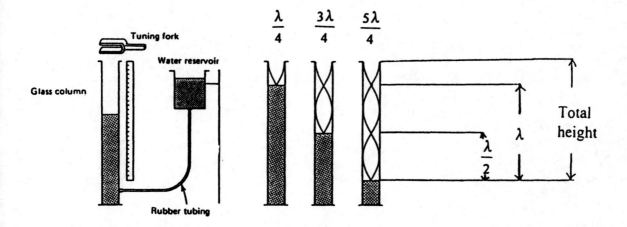

5. Strike the tuning fork again and continue to lower the water level until another resonance is observed. Determine the position of all resonances.

6. Each resonance is given by an odd number of quarter wavelengths (1/4, 3/4, 5/4, etc.). From the position of each resonance found in step 5, record the wavelength. Then determine the velocity of sound.

$$\lambda_1 = 4d_1 \qquad \lambda_2 = 4d_2/3 \qquad \lambda_3 = 4d_2/5$$

Wavelength	Resonance	Height of air column	λ	$v_{measured}$	v_{actual}	% error
$\lambda/4$	1	m	m	m/s	m/s	
$3\lambda/4$	2	m	m	m/s	m/s	
$5\lambda/4$	3	m	m	m/s	m/s	

$$v_{calc} = 331 + 0.61T \qquad \qquad \% \text{ error} = \frac{v_{measured} - v_{calc}}{v_{calc}} \times 100$$

Question:

1. What are the possible sources of error in this experiment?

ELECTROSTATICS

Experiment
22

Name _____

Date _____

Period _____

Lab Partner _____

Purpose:

The purpose of this lab is to develop familiarity with some of the characteristics of static electricity.

Equipment:

Electrostatic needle
Electroscope
Electrophorus
Proof plane
Wool and silk patches
Glass and hard rubber rods
Pith balls and string

Procedure:

In this lab you will not be given a set of specified instructions to follow. You will be expected to take the equipment and investigate static electricity on your own. Try to produce both positive and negative charges by conduction (touching) and induction (only bringing near).

Caution: Use extreme care when attempting to charge a gold-leaf electroscope. Too much charge will tear the leaf off its post.

Try to produce the above effects on your own. If you have great difficulty, some suggestions are given below.

A negative charge is placed on a rubber rod by rubbing it with fur or wool.

A positive charge is placed on a glass rod by rubbing it with silk.

The electrophorus-charging by induction

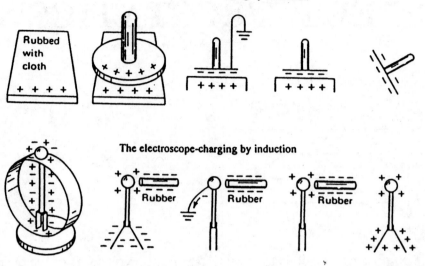

The electroscope-charging by induction

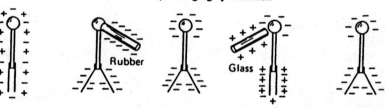

The electroscope-charging by induction

Questions:

1. Why does the gold leaf pull away from the stem of the electroscope when a charge is placed on it?

2. What behavior indicates that an object is electrified?

3. Is this behavior selective, like magnetism, or is it shown toward all bodies?

4. How did you show that there are two kinds of static electricity?

5. What did you discover about like charges? Unlike charges?

6. Name two different ways to place a charge on a body.

METERS AND OHM'S LAW

Experiment

23

Name _____

Date _____

Period _____

Lab Partner _____

Purpose:

The purpose of this lab is to apply Ohm's law to a simple circuit and to study the ammeter and voltmeter.

Equipment:

Two no. 6 (1.5-V) dry cells
One 6-V dry cell
Connector wires
Rheostat, approximately 0-25 Ω
dc voltmeter (0-7.5 V)
dc ammeter (0-1 A)
Switch
Fuse block
1-A fuse
Coil

Caution:

Ac and dc meters are constructed differently and may not be interchanged. Be certain you have dc meters for this lab. Meters are delicate instruments and must be handled with extreme care. In using an instrument with more than one range (for example, an ammeter with ranges of 0—1 A and 0—10 A), always start with the highest range. If no reading appears on the meter, then switch to a lower scale. Voltmeters measure voltage drop and must be connected in parallel with the circuit component. Ammeters measure current in a circuit and must be connected in series in the circuit.

Procedure:

dc Meters:

1. Draw the circuit diagram for the setup shown below in the space provided:

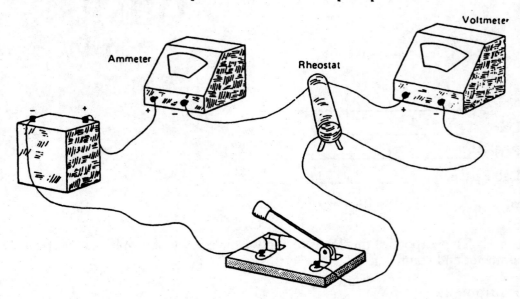

2. Draw the circuit diagram for an ac source or generator and three lamps in parallel.

3. Draw the circuit diagram for a 1.5-V cell in series with a 1.5 - Ω resistor and two 3-Ω resistors in parallel. **Find the total resistance, R_T; the main current, I_T; and branch currents.**

4. Set up the circuits shown below. Your instructor will check your work.

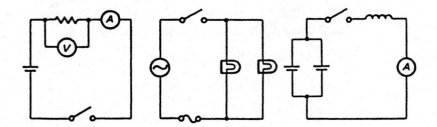

5. Set up the circuit shown below using the 1.5-V dry cell. Be careful to observe the polarities of the meters.

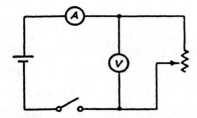

6. Close the switch and adjust the rheostat so that the ammeter reads 0.10 A. Record the ammeter and voltmeter readings in the data table on the next page.

7. Find R by calculating:

$$R = \frac{V}{I}$$

8. Repeat steps 6 and 7, adjusting the rheostat for currents of 0.30 A and 0.50 A.

9. Repeat the above procedure using a 6-V dry cell and complete the data table.

10. Plot a V versus I graph of the data and determine the slope. The slope gives the resistance.

11. Find the percentage difference between the slope of your curve and the measured value of the resistance.

Emf	Trial	I	V	Rco/c	Rmeas
1.5V	1	0.10 A	V	Ω	
	2	0.30 A	V	Ω	
	3	0.50 A	V	Ω	
6V	1	0.10 A	V	Ω	
	2	0.30 A	V	Ω	
	3	0.50 A	V	Ω	

Questions:

1. Does the value of R change when you replace the 1.5-V source with the 6-V source?

2. How does the value of V change as I increases from 0.10 A to 0.50 A?

3. Is the value of V reasonably close to the value of the emf?

4. List three sources of possible error in this experiment.

5. Distinguish between "emf" and "voltage."

6. What is "resistivity"?

RESISTANCE IN A WIRE

Experiment 24

Name _____

Date _____

Period _____

Lab Partner_____

Purpose:

The purpose of this lab is to determine the relationship between the electrical resistance of a wire and its length, its cross-sectional area, and its material.

Equipment:

New dry cell
Ammeter
Voltmeter
Connecting wires
Switch
100 cm No. 24 manganin wire
100 cm No. 24 nickel silver wire
100 cm No. 36 nickel silver wire
150 cm No. 24 nickel silver wire
200 cm No. 24 nickel silver wire

Procedure:

1. Connect the apparatus as indicated in the diagram on the next page using the 100 cm No. 24 manganin wire. Remember that the ammeter is connected in series and the voltmeter is connected in parallel.

2. Close the switch and read the ammeter and voltmeter. Record your reading in the table.

3. Repeat steps I and 2 for the 100 cm No. 24 nickel silver wire, 100 cm No. 36 nickel silver wire, 150 cm No. 24 nickel silver wire, and the 200 cm No. 24 nickel silver wire.

4. Calculate the resistance for all trials using Ohm's law: $V = IR$.

Manganin or nickel silver wire

Trial	Kind of Wire	Length	E	I	R_{meas}	$R_{\text{cal}}=\rho l/A$	$E/I = R_c$	% diff. error
1	No. 24 manganin	100 cm	V	A			Ω	
2	No. 24 nickel silver	100 cm	V	A			Ω	
3	No. 36 nickel silver	100 cm	V	A			Ω	
4	No. 24 nickel silver	150 cm	V	A			Ω	
5	No. 24 nickel silver	200 cm	V	A			Ω	

Questions:

1. For trials 1 and 2: How does the resistance depend upon the material of the wire?

2. For trials 2 and 3: How does the cross-sectional area of the wire affect the resistance?

3. For trials 2, 4, and 5: How does the length affect the resistance of a wire?

4. What is the difference between "Resistance" and "Resistivity?" Which is constant for a given material?

SERIES AND PARALLEL CIRCUITS

Experiment **25**

Name _____

Date _____

Period _____

Lab Partner _____

Purpose:

The purpose of this lab is to study the relationships of current, resistance, and voltage in simple circuits.

Equipment:

dc ammeter (0 - 1 A)
dc voltmeter (0 - 3 V)
Six resistors
Connector wires
Switch
1.5-V or 6-V dry cell

Procedure:

1. Set up the circuit shown using three resistors, each having the same resistance.

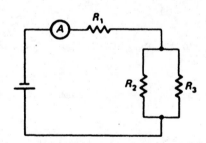

2. Using the voltmeter, measure the voltage drop across each load (resistor). Record all readings in the data table.

3. Record the ammeter reading.

4. Compute the current through R_1 by using V_1, R_1, and Ohm's law $(V_1 = I_1 R_1)$.

5. Compute the equivalent resistance of R_2 and R_3.

6. Repeat steps 1-5 with different resistors, each having different values.

	Trial 1	Trial 2
R_1 (mess)	Ω	Ω
R_2 (mess)	Ω	Ω
R_3 (mess)	Ω	Ω
V_1 (mess)	V	V
V_2 (mess)	V	V
V_3 (mess)	V	V
I (mess)	A	A
I Computed	A	A
R compute equivalent	Ω	Ω

Using the ammeter reading and the calculated equivalent resistance, compute the voltage drop across R_2 and R_3. Compare the computed values to the measured values from step 2.

Trial	V_2 Computed	V_3 Computed
1		
2		

Questions:

1. Add $V_1 + V_2$. Does this sum equal the emf of the cell?

2. Using R_2, R_3, V_2, and V_3, compute I_2 and I_3. Does the sum Of $I_2 + I_3$ equal the ammeter reading?

3. What did you learn about the relationship between the voltage drops in the circuit and the emf?

4. Verify Ohm's law using the sum of the voltage drops, the total current and the equivalent resistance of the circuit.

5. Distinguish between "EMF" and "Voltage".

6. With a given set of resistances, how would you connect to obtain (1) a smaller equivalent resistance and (2) a larger equivalent resistance?

7. Answer question six (6) if resistances are replaced by capacities.

TEMPERATURE COEFFICIENT OF RESISTANCE

Experiment **26**

Name _____

Date _____

Period _____

Lab Partner _____

Purpose:

The purpose of this experiment is to measure the variation of resistance with temperature of a metal.

Equipment:

Temperature-coefficient coil
Thermistor
Thermometer
Ohmmeter (electronic)

Theory:

The variation of resistance with temperature can be expressed as an infinite series such as

$$R_T = R_0 + R_1 T + R_2 T^2 + \ldots \ldots$$

where R_0 is the resistance at temperature $T = 0°$ C and R_1, R_2... are resistances at other temperatures T_1, T_2.....

In case of metals, R_2, R_3, and higher order coefficients are very small. For small changes of temperature, the terms in T^2 and higher powers of T can be neglected. R_1 is nearly constant and is written as a R_0, where a is the temperature coefficient of resistance. Then the above equation becomes

$$R_T = R_0 + R_0 T = R_0(1 + aT).$$

For metals, an increase in temperature causes a slight increase in the number of collisions of electrons with lattice defects and impurities. This increase in collisions decreases the current flow. The resistance is given by $R = \dfrac{V}{I}$, where V is the voltage and I is the current. When current decreases, the resistance increases.

Procedure:

1. Use cool water from the drinking fountain to fill the vessel to the least possible level that will completely cover the temperature-coefficient coil. The initial water temperature should be less than 15°C, preferably around 10°C.

2. Place the heating vessel in the fiber ring of the tripod and connect the heating unit with the card supplied to a 120-V AC outlet.

3. Insert the thermometer in the center hole of the cover that holds the coil to be tested.

4. Connect the electronic thermometer to the coil.

5. Measure the temperature of water using the 10 ohm scale, and determine the coil resistance R_T for this temperature of water.

6. Heat the water so that the temperature will increase 8° or 10°. Stir the water during the heating process and after the heater is turned off until the temperature stops rising. Be sure that the thermometer is not resting on the heating element. Carefully and quickly read both the thermometer and ohm meter. Record the temperature T and resistance R.

7. Repeat step 6 until the temperature reaches approximately 90°C.

8. Plot the values of resistance R_T against T on a linear graph. Draw the best fit straight line through data points. Determine the intercept and the slope. The intercept of the line is R_0 and the slope is $R_0 a$. Find a, the temperature coefficient of resistance. Compare your value with the book value and calculate the percentage error.

POTENTIOMETER METHOD OF EMF

Name _____

Date _____

Period _____

Lab Partner _____

Purpose:

The purpose of this experiment is to study the principles of the potentiometer as a method of measuring emf, and to measure the emf of several batteries.

Equipment:

>Slide-wire potentiometer with contact key
>Switch
>Galvanometer
>Standard cell
>Several unknown dry cell batteries
>Lab-volt
>Protective switch

Theory:

Potentiometer refers to a special type of resistor with three terminals. Two terminals are at opposite ends of a resistor, and the third terminal may be moved to contact this resistor at any point from one end to the other. Potentiometers are very commonly used for TV or radio volume control.

A slide wire potentiometer is used in this experiment to measure the emf of batteries. It consists of a resistance wire one meter long and a tap key K, which can contact the wire at any point along its length.

The emf of a battery is the potential difference between its terminals when **no** current is flowing. This emf cannot be measured with a voltmeter because the battery current needed to drive the voltmeter produces a voltage drop, IR, across the battery's internal resistance. The voltmeter reading is less than the emf, i.e.,

$$V = E - IR$$

However, in a circuit like that shown in **Figure 1**, the potential drops gradually along the slide wire from A to B. A location, K, can be found where the potential difference between A and K will match the emf of the battery. Since the potentials match and oppose each other, no current flows through the battery. The emf equals the potential difference between A and K, and the problem then is to determine the potential difference from A to K.

The way this problem is solved is that (a) to connect a standard cell with known emf, E_s, into the circuit. Notice that the Lab-volt creates a potential on the wire which opposes the emf of the standard cell. The movable contact K is adjusted until no current flows from E_s, as indicated by no deflection on the galvanometer. Let R_s be the resistance from A to K, and L_s be the distance from A to K when using the standard cell. (b)Then the same procedure is followed with the unknown emf E_u, calling R_u the resistance and L_u the distance from A to K with the unknown battery in the circuit. We can then write

$$E_s = IR_s \text{ and } E_u = IR_u$$

where the current in the wire, I, is the same in both cases since it is produced by the Lab-volt. If we divide one equation by the other, the result is

$$E_u/E_s = R_u/R_s$$

The resistance of length L of wire with cross-sectional area A is $\rho L/A$, where ρ is the resistivity of the wire. Substituting this we have

$$E_u/E_s = R_u/R_s = (\rho L_u/A) / (\rho L_s/A) = L_u/L_s \qquad (1)$$

We know that L_u, L_s, and E_s are the measurements necessary to determine that unknown emf, E_u.

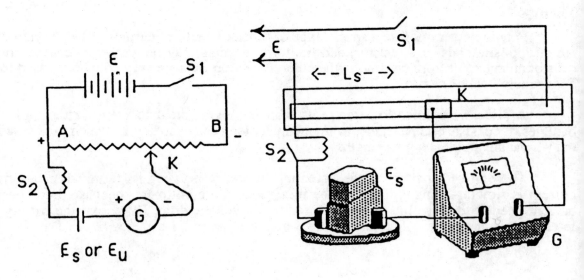

Figure 1   Figure 2

98

Procedure:

1. Connect the slide-wire potentiometer as shown in Figure 2. The Lab-volt must provide sufficient potential to match the unknown emf's. Three volts is sufficient for this experiment. The exact value is unimportant: what is important is that this potential remains unchanged during the experiment. A voltmeter across the Lab-volt terminals can be used to monitor this voltage and verify that the output potential remains the same whenever switch S 1 is in the CLOSED position. (When S_1 is in the OPEN position, the Lab-volt output will change, but this is of no concern.) <u>The instructor will insert the standard cell in your circuit. You can then measure the value for Ls, which will be used throughout the experiment</u>.

2. Replace the standard cell with a battery of unknown emf.

3. When you are ready to take a reading of an unknown, close switch S I and tap key K with a sharp tap, making the contact as short as possible, and note the direction of deflection of the galvanometer first near the positive end of the slide wire and then near the negative end. The two deflections should be in opposite directions. If not, immediately open the switch and check the circuit for proper wiring and connections. Consult the instructor if the problem cannot be calculated.

4. Adjust the tap key K until there is no deflection in the galvanometer. Close switch S_2, which will make the galvanometer more sensitive, and adjust the tap key K until there is no deflection once again. The length Lg is measured from the positive end. The emf of the unknown can now be calculated.

5. Repeat steps 2, 3, and 4 for the two other unknown batteries.

Questions:

1. Why should you not hold your fingers across the terminals of an unknown when taking a reading for L_U?

2. The voltage of the Lab-volt was not recorded or used in calculations. Why?

3. The voltage of an unknown cell as indicated by a voltmeter should be less than that found by potentiometer. Explain why.

4. A battery with a emf of 1.52 volts is connected to a voltmeter which records 1.48 volts. The resistance of voltmeter is 200 ohms. Calculate the internal resistance of the battery.

LEAD STORAGE BATTERY

Experiment **28**

Name _____

Date _____

Period _____

Lab Partner _____

Purpose:

The purpose of this lab is to learn how the basic lead storage cell works, that is, how it becomes charged and then releases its energy.

Equipment:

Glass tumbler with porcelain top, with two clamps for holding lead plates, with two binding posts for wire connections
Two lead plates
Three 1.5-V dry cells
Voltmeter
Connector wires
Sandpaper
Electric bell
Sulfuric acid (concentrated)—mix 1 part acid to 6 parts water

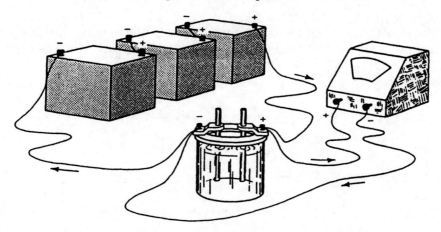

Caution: Be very careful with the sulfuric acid solution; very small amounts will ruin your clothing.

Procedure:

        Clean the lead plates with sandpaper until they are bright and any corrosion is removed. Place the plates in the porcelain holder; then insert them in the acid solution. Connect the voltmeter. What is the reading? _____V

Explain.

        Next, connect three dry cells in series to the lead plates as in the diagram on page 101. Connect the voltmeter in parallel across the lead storage cell so that the (+) pole of the lead cell is connected to the (+) on the voltmeter. Allow current to flow through the lead cell for five minutes.

Describe what you observe in the cell.

Examine the plates. What is the color of each plate?

Disconnect the dry cells. Read the voltmeter. _____V

Immediately connect the lead cell to an electric bell. Describe the voltmeter reading as the bell rings and the lead cell is discharging.

Next connect the negative and positive terminals with a wire, thus short-circuiting the cell. Wait about three minutes; then connect the voltmeter and record its reading. _____V

Explain.

Describe the color of the plates.

Is the positive plate (when charging) positive or negative when discharging?

MAKING A DRY CELL

Experiment 29

Name _____

Date _____

Period _____

Lab Partner _____

Purpose:

The purpose of this lab is to make a dry cell from a kit such as the Burgess Battery dry cell kit and to test the dry cell for current and voltage in a circuit.

Equipment:

       Dry cell kit consisting of-
          Capped carbon rod
          Zinc can
          Separator paper
          Mix (manganese dioxide, carbon, and electrolyte)
       Voltmeter: 0-5 or 0- 10 V
       Ammeter: 0-20 or 0-25 A
       Flashlight bulbs
       Doorbell buzzer

Procedure:

The kit contains the directions for assembling the dry cell. Test the assembled dry cell for voltage and current in a circuit. Then set up simple circuits using one, two, or more cells in series and parallel to test current and voltage. Try flashlight bulbs, doorbell buzzers, and the like, in the circuit and observe its operation.

If time permits, make other dry cells using less mix, changing electrolytes, mixing electrolytes, etc.

CELLS IN SERIES AND PARALLEL

Name _____

Date _____

Period _____

Lab Partner _____

Purpose:

The purpose of this lab is to show how cells may be grouped to supply various currents and voltages and to measure the internal resistance of a battery.

Equipment:

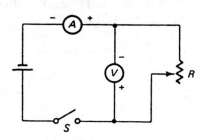

Three No. 6 (1.5-V) dry cells
dc ammeter (0-500 mA)
Rheostat, approximately 25 W
dc voltmeter (0-7.5 V)
Connectors
Connector wire
Two knife switches

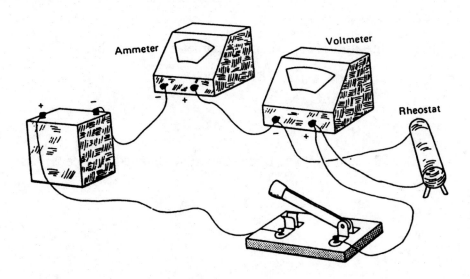

Procedure:

Cells in Series

1. Using just one dry cell, set up the circuit shown on page 105. Adjust the rheostat for maximum resistance and close the switch. Read both meters, but do not record data.

2. Adjust the rheostat so the ammeter reads precisely 0.5 A. Record this value and the voltmeter reading in the data table.

3. Place a second dry cell in series with the one already in the circuit. (Connect the negative terminal of the first to the positive terminal of the second.) Close the switch and record both meter readings.

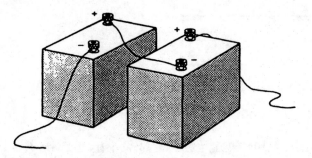

4. Place a third dry cell in series with the others. Close the switch and record the meter readings.

Number of Cells	Ammeter Reading (I)	Voltmeter Reading (V)
1	A	V
2	A	V
3	A	V

Questions:

(a) Does the current in the circuit (ammeter reading) change?

(b) Does the emf supplied by the battery (combination of cells) change?

Cells in Parallel

5. Set up a circuit with only one dry cell as in step 1, but this time adjust the rheostat to obtain an ammeter reading of 0.25 A. Record the meter readings.

6. Place a second dry cell in parallel with the first one in the circuit. (Connect positive (+) terminals of the two cells together and the negative (-) terminals together.) Close the switch and record the meter readings.

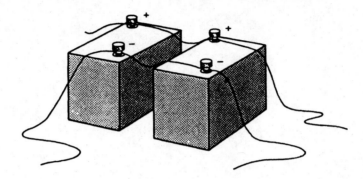

Place a third dry cell in parallel with the first two. Close the switch and record the meter readings.

Cells in Parallel

Number of Cells	Ammeter Reading (I)	Voltmeter Reading (V)
1	A	V
2	A	V
3	A	V

7. Discuss the advantage/disadvantage of connecting cells in series and then in parallel.

Questions:

(a) Does the current in the circuit change?

(b) Does the emf supplied by the battery (combination of cells) change?

ELECTRICAL EQUIVALENT OF HEAT

Experiment

31

Name _____

Date _____

Period _____

Lab Partner _____

Purpose:

The purpose of this lab is to make some measurements to compare the amount of heat energy absorbed by water to the electrical energy expended by a heating coil.

Equipment:

Styrofoam cup (calorimeter)
Heating coil
dc voltmeter, dc ammeter (0-5 A)
Clock
Platform balance
Thermometer
Storage battery

Procedure:

1. Weigh the empty styrofoam cup.

2. Fill the cup 2/3 full with water 15°C below room temperature.

3. Weigh the cup and water.

4. Connect the circuit as shown in the diagram on the next page.

5. Stir water thoroughly and record initial temperature.

6. Close switch and heat for two minutes. Record the voltmeter and ammeter readings.

7. After two minutes, open, switch, remove the coil, stir thoroughly, and record the final temperature of the water.

8. Carry out computations (for conversion from joules to calories use 1 cal = 4.19 J).

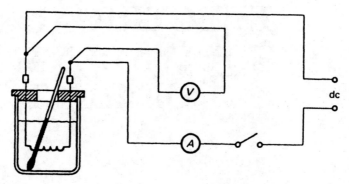

	Trial 1		Trial 2	
Mass of calorimeter, m_c		g		g
Mass of water, m_w		g		g
Initial temperature of water, T_i		°C		°C
Final temperature of water, T_f		°C		°C
Rise in water temperature, ΔT		°C		°C
Specific heat of water, C_w		cal/g°C		cal/g°C
Heat absorbed by water, $Q = m_w C_w \Delta T$		cal		cal
Voltage		V		V
Current		A		A
Time		s		s
Electrical energy input $J = VIt$		J		J
Electrical energy input		cal		cal

Questions:

1. How does the electrical energy in calories compare with the heat energy absorbed?

2. Why should the temperature of the water be below room temperature to begin the experiment?

3. Why should the temperatures be taken promptly after the coil is removed?

4. List three possible sources of error in the experiment.

THE MAGNETIC FIELD

Experiment 32

Name _____

Date _____

Period _____

Lab Partner _____

Purpose:

The purpose of this lab is to determine the direction of the magnetic field near a current carrying wire. You will determine the direction using a magnetic compass.

Equipment:

Storage battery or dc power supply
Rheostat
dc Ammeter (0-5 A)
Wire compass

Procedure:

1. Determine which end of the compass needle points to the north pole.

2. Wire the equipment as shown in the diagram below. Be sure to keep all sections of wire, except the long straight portion, at least 50 cm away from the compass.

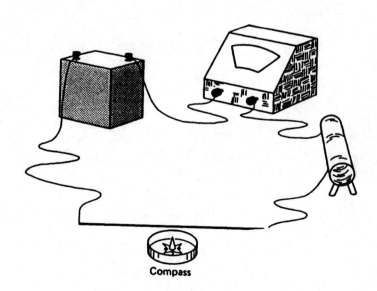

Compass

3. Adjust the rheostat for a current of about 5 A.

4. Slide the compass under the wire and indicate the direction to which it points in the diagram.

5. Now hold the compass above the wire and indicate the direction to which it points.

6. Disconnect one lead from the battery. What happens to the compass needle?

7. Reverse the connections to the battery and repeat steps 4 and 5.

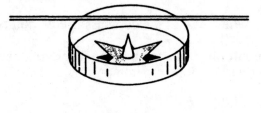

8. Suspend the long section of wire from a support stand as shown on page 113.

9. Using the right-hand rule, predict the direction of the field about the wire. Record the direction in the diagram.

Current out

112

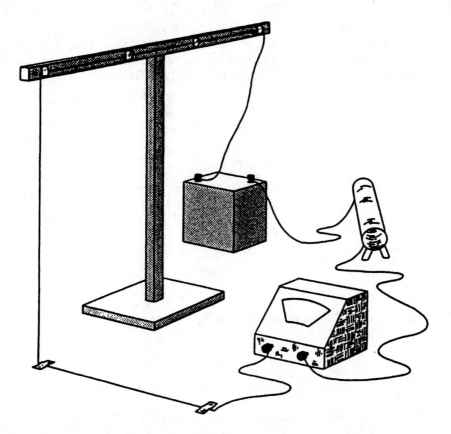

10. Adjust the rheostat for a current of about 5 A. Move the compass around the wire in a horizontal plane and record the direction of the field in the diagram.

Current out

How does the direction compare to that predicted in step 9?

11. Reverse the battery connections and repeat steps 9 and 10.

Current in

Current in

12. Wind the wire into a coil of three or four turns of about 1-1/2 in. in diameter. Move the compass in and around the coil and record the directions of the magnetic field at the points in the diagram marked with an x.

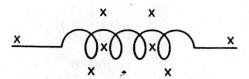

13. Reverse the battery connections and repeat step 12.

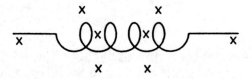

INDUCTION

Name _____

Date _____

Period _____

Lab Partner _____

Purpose:

The purpose of this lab is to study electromagnetic induction. We will study the factors that determine the size and direction of induced currents, and see how they apply to the construction of generators.

Equipment:

Student induction coils with iron core
Permanent magnet (horseshoe)
1.5-V dry cell
Galvanometer (zero center)
Switch
Rheostat, approximately 10 ohms
Compass
Generator coil

Equipment Note: If the above coil is unavailable, wrap about eighty turns of No. 28 insulated copper wire on a test tube and about forty turns of No. 22 wire on a smaller tube. Secure the coils so the small tube may be slipped inside the large one.

Procedure:

Part 1: Permanent magnet induction

1. Connect a large coil to the galvanometer. Thrust the N pole of a permanent magnet through the coil and observe the galvanometer. Result?

2. Leave the magnet in the coil and do not move it. Is there any meter deflection?

3. Pull the magnet out of the coil. Any deflection?

4. Repeat the procedure in steps 1-3 while varying the speed with which you thrust the coil. What can you conclude from these experiments?

5. Turn the magnet around and thrust the S pole into the coil. Result?

Part 2: The electromagnet

1. Take the large coil and connect it in the circuit shown below.

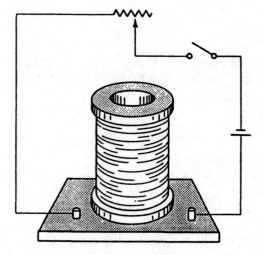

2. Bring a compass near the coil (with the switch closed). Is there a magnetic field around the coil?

3. Try to pick up a small nail with the coil. Result?

4. Insert an iron core into the coil. Again bring a compass near the coil. Result?

5. With the core in the coil, again try to pick up the nail. Result?

6. What can you conclude about the effect of the iron core on the strength of the magnetic field?

Part 3: Induction with an electromagnet

1. Set up the circuit shown. The source is connected to the primary coil and the galvanometer to the secondary. The primary and secondary coils are not wired together.

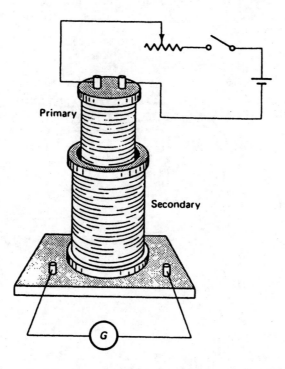

2. Slip the small coil into the large coil and observe the galvanometer deflection:

   (a) When the circuit is being closed.

   (b) After the circuit has been closed a few seconds.

   (c) When the circuit is being opened again.

   (d) When the current is increased (by reducing the resistance of the rheostat).

   (e) When the current flow is reversed (by changing the leads on the battery terminals).

   (f) When an iron core is placed in the primary coil. Caution: Insert the core slowly and adjust the rheostat as necessary to protect the galvanometer.

<u>Part 4</u>: **The generator**

1. Make a coil of about eighteen turns such as the one shown, and connect it (with long leads) to the galvanometer. Make the coil small enough to fit sideways in the horseshoe magnet.

2. Give the coil a sudden twist in the magnet and observe the meter deflection, if any.

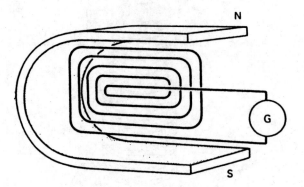

3. Continue to rotate the coil in the magnet. Is there any part of the rotation during which there is no deflection on the galvanometer?

4. In one rotation, is the current always in the same direction?

5. Rotate the loop in the opposite direction. Result?

6. Does your coil generate ac or dc current?

7. Plot a graph of current versus time.

dc   <small>Experiment</small>
MOTORS   34

Name _____

Date _____

Period _____

Lab Partner _____

Purpose:

The purpose of this lab is to study the construction and operation of dc motors.

Equipment:

- St. Louis motor with a dc rotor
- Field winding (electromagnet)
- Dry cell or storage cell
- Compass
- Ammeter (0-1.5A)

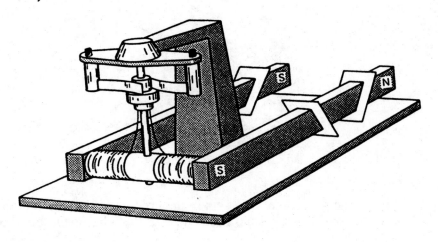

Procedure:

1. Install the dc rotor in the motor.

2. Remove the bar magnets.

3. Connect the dry cell to the rotor.

4. Use the compass to check the polarity of the rotor in several different positions as you slowly turn it through one revolution.

5. Record the polarity at 30° intervals in **diagram 1**.

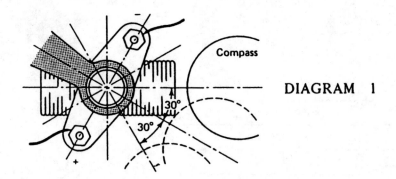

DIAGRAM 1

6. What causes the change in polarity of the rotor?

7. Reverse the dry cell connections to the rotor.

8. Test and record the polarity of the rotor through one rotation in **diagram 2**. How does the polarity compare with that found in steps 4 and 5?

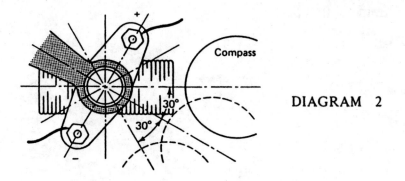

DIAGRAM 2

9. Place the permanent magnets in the holders so that the S pole of one magnet is on one side of the rotor and the N pole of the other is on the opposite side.

10. Reverse the connections between the dry cell and the rotor.

11. Give the armature a push to start it rotating. Record in diagram 1 the polarity of the permanent magnets and the direction of rotation.

12. Adjust the brushes to get the maximum rotational speed.

13. Reverse the connections to the rotor and record in diagram 2 the direction of rotation.

14. Remove the bar magnets and install the electromagnet. Connect the rotor and electromagnet in series with the dry cell as shown on the next page.

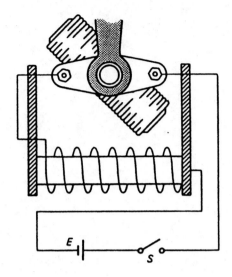

15. Start the motor and note the direction of rotation. Reverse the dry cell connections. Does the direction of rotation change?

Why?

16. Reverse the connections to the magnet only. Does the direction of rotation change?

17. Replace the electromagnet with the bar magnets. Connect a dc ammeter in series with the rotor. Start the motor and note the meter reading as the rotor comes up to speed. What do you observe?

ac Experiment
MOTORS 35

Name _____

Date _____

Period _____

Lab Partner _____

Purpose:

The purpose of this lab is to study the construction and operation of ac motors.

Equipment:

St. Louis motor with ac and dc rotors
Field winding (electromagnet)
Dry cell or storage cell
6-V transformer
Electric motor

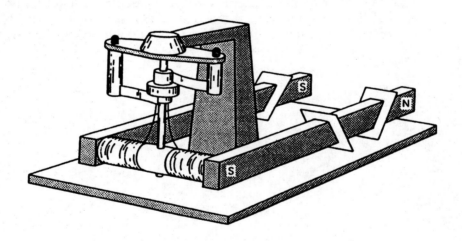

Procedure:

1. Install the dc rotor (with commutator) in the motor.

2. Connect the field magnet and rotor in series with the 6-V dc source.

3. Start the motor and note the direction of rotation.

4. Reverse the field magnet connections and note the direction of rotation. Has it changed?

5. How does this motor (called a universal motor) compare with the dc motors studied in the dc motors lab?

6. Install the ac rotor and the bar magnets.

7. Connect the rotor to the 6-V ac supply.

8. Try to start the motor by hand. Why won't it start?

9. Use the electric motor supplied by your instructor to get the St. Louis motor up to 3600 rpm and then let it run. This is a synchronous motor of the type used in clocks. Why is the speed 3600 rpm (60 revolutions per second)?

LAMPS IN SERIES AND PARALLEL

Experiment **36**

Name _____

Date _____

Period _____

Lab Partner _____

Purpose:

The purpose of this lab is to study series and parallel wiring of ac circuits.

Equipment:

Three socket lamp boards
Fuse block and 6-A fuses
Split-line cord and plug
Connector wires
Three 60-watt light bulbs
ac voltmeter (0 - 150 V)
ac milliammeter (0 - 500 mA)
ac ammeter (0 - 3 A)

<u>Caution</u>: Make certain the plug into the ac source has been removed before changing any connections. You may otherwise receive a dangerous shock.

Procedure:

<u>Part 1</u>: **Series**

1. Set up the circuit shown. <u>Do not</u> plug in the power line until your setup has been checked by the instructor.

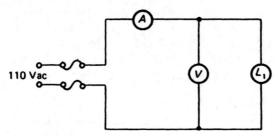

2. Record the ammeter and voltmeter readings for L_1 in the data table on the next page.

3. Substitute L_2 for L_1 and record the meter readings.

4. Substitute L_3 for L_2 and record the meter readings.

5. Remove L_3 from the circuit and connect L_1 and L_2 in series in the circuit. Record the meter readings.

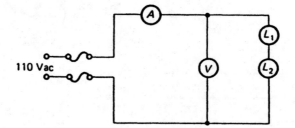

6. Place L_3 in series with L_1 and L_2. Record all meter readings.

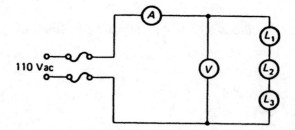

7. How does the intensity of the light given off by each lamp in series compare with the intensity of L_1 alone?

8. Unscrew one of the lamps. What happens?

9. Using the data obtained in steps 1-6, calculate the resistance of each lamp and complete the series part of the data table.

Part 2: Parallel

1. Set up the circuit shown with L_1 and L_2 in parallel. Have your instructor check your circuit before proceeding.

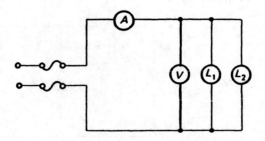

2. Record the meter readings.

3. Connect L_3 in parallel with L_1 and L_2.

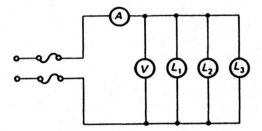

4. Record the meter readings.

5. How does the intensity of light given off by each lamp in parallel compare with the intensity of L_1 alone?

6. Loosen one of the lamps. What happens?

7. Using the data obtained in steps 1-4, calculate the equivalent resistance of each parallel circuit.

	Lamps	V	I	R
	L_1	V	A	Ω
	L_2	V	A	Ω
	L_3	V	A	Ω
Series	$L_1 + L_2$	V	A	Ω
	$L_1 + L_2 + L_3$	V	A	Ω
Parallel	$L_1 + L_2$	V	A	Ω
	$L_1 + L_2 + L_3$	V	A	Ω

TRANSFORMERS Experiment
37

Name _____

Date _____

Period _____

Lab Partner _____

Purpose:

The purpose of this lab is to show how voltage may be changed by a transformer.

Equipment:

Closed-core transformer with removable coils of at least 100 turns (commercial or homemade)
At least four coils of different numbers of turns
ac voltmeter
Fuse block and 10-A fuses

Procedure:

1. Set up the circuit shown using an ac source provided by your instructor or the 110-V lighting outlet. <u>Caution:</u> Do not plug in the circuit until it has been checked by your instructor.

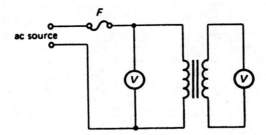

2. Vary the coils used and record in the table on page 130 the number of turns and voltage for each coil.

3. Compute the ratios N_s/N_p and V_s/V_p. Are they nearly the same? Should they be? Plot a graph of V_s and N_s when N_p is held constant. Can you predict a value for V_s if given a number of N_s? What is the shape of the graph?

N_p	N_s	V_p	V_s
		V	V
		V	V
		V	V
		V .	V
		V	V
		V	V

Calculated results

N_s/N_p	V_s/V_p

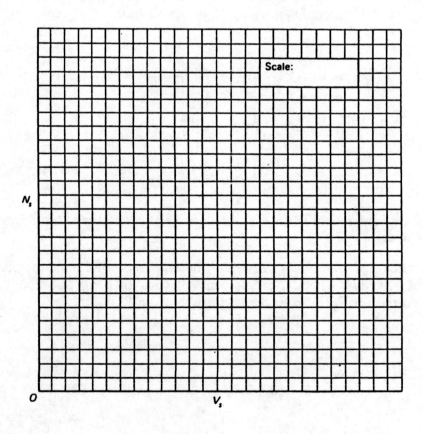

CAPACITANCE

Experiment **38**

Name _____

Date _____

Period _____

Lab Partner _____

Purpose:

The purpose of this lab is to study the charge-storing ability of capacitors and their effect on ac circuits.

Equipment:

Knife switch, DPST
Four dry cells
dc milliammeter (0-50 mA)
dc voltmeter
6-V lamp (No. 40 miniature screw base)
Lamp base
25-V capacitors (one 25 μF and one 50 μF)
6-V ac source
ac milliammeter (0-500 mA)
ac voltmeter

Procedure:

CHARGING OF CAPACITORS:

1. Make a 6-V dc source by connecting the four dry cells in series.

2. Wire the circuit shown.

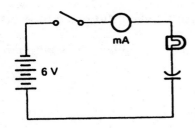

3. Close the switch while observing the ammeter. Was there any movement of the meter needle?

What was happening to the capacitor?

Did the lamp light?

4. Connect the voltmeter across the capacitor and open the switch. What happens?

You are observing the discharge of the capacitor through the internal resistance of the voltmeter.

5. Replace the 25-μF capacitor with the 50-μF capacitor. Repeat the above observations and comment on any differences.

CAPACITORS IN ac CIRCUITS:

1. Replace the dc instruments with the ac instruments and replace the battery with the 6-V ac source.

2. Close the switch and record the current with the 50-μF capacitor.
   I = _____ mA.

3. Measure and record the voltage across the capacitor and the lamp.

4. Open the switch and replace the 50-μF capacitor with the 25-μF capacitor.

5. Close the switch and record the current and the voltages across the capacitor and the lamp.

6. Do the necessary calculations to fill in the following table.

7. How does the calculated current compare with the measured current?

Capacitance	μF	μF
Voltage Across Capacitor (V_C)	V	V
I	mA	mA
Voltage Across Lamp (V_L)	V	V
Total Voltage $V_T = V_L + V_C$	V	V
$R = \dfrac{V_L}{I}$	Ω	Ω
X_L	Ω	Ω
Z	Ω	Ω
$I = \dfrac{V_T}{Z}$	mA	mA

INDUCTANCE   Experiment
39

Name _____

Date _____

Period _____

Lab Partner _____

Purpose:

The purpose of this lab is to study the effect of inductance on ac circuits.

Equipment:

6-V storage battery
Switch
Primary and secondary coil set with iron core
6-V carbon filament lamp socket
dc ammeter (0-3 A)
dc milliammeter (0-100 mA)
dc voltmeter
6-V ac source
Tubular rheostat (10 Ω)
ac ammeter (0-3 A)
ac milliammeter (0-50 mA)
ac voltmeter

Procedure:

INDUCTOR IN dc CIRCUIT:

1. Wire the circuit shown. Use the primary coil as the inductor.

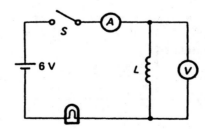

2. Close the switch and record the voltage across the coil and the current flow. Determine the resistance of the coil. Substitute the secondary coil for the primary coil. Repeat for the secondary coil using the milliammeter in place of the ammeter.

	V	I	R
Primary	V	A	Ω
Secondary	V	A	Ω

INDUCTOR IN ac CIRCUIT:

3. Replace the dc instruments with the ac instruments.

4. Replace the dc source with the ac source. Use the secondary coil.

5. Observe the lamp intensity while inserting and withdrawing the iron core. Result?

FINDING THE INDUCTANCE OF THE COIL:

6. Remove the lamp from the circuit and replace with the rheostat. Close the switch and adjust the rheostat to give a convenient current reading. Record the current and voltage.

7. Do the necessary calculations to complete the table.

I	V_L	R_L from preceding table	Z	$X_L = Z - R_L$	L
mA	V	Ω	Ω	Ω	H

SEMICONDUCTOR DIODES

Experiment

40

Name _____

Date _____

Period _____

Lab Partner _____

Purpose:

The purpose of this lab is to investigate the rectification characteristics of semiconductor diodes.

Equipment:

Oscilloscope
Sine-wave generator
Semiconductor diodes
1 kΩ resistor
Connecting wires

Procedure:

HALF-WAVE RECTIFICATION:

1. Set up the circuit shown.

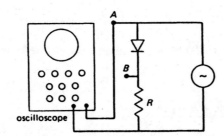

2. Set the sine-wave generator frequency to 100 Hz. Adjust the amplitude of the waveform to be approximately one-half the screen height. Measure the peak-to-peak ac voltage at the signal generator by connecting the oscilloscope vertical input to position A. Sketch the waveform.

3. Measure the voltage drop across the resistor by connecting the vertical input to point B. Sketch the waveform.

4. Explain the difference between the waveforms measured at A and B.

FULL WAVE RECTIFICATION:

1. Set up the bridge rectification circuit shown below.

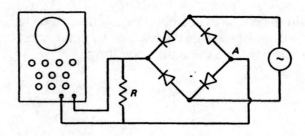

2. Sketch the rectified waveform.

3. Explain the observed waveform.

FILTERING:

1. Add the filter circuit shown below.

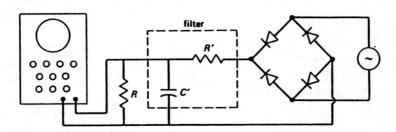

2. Sketch the observed waveform.

3. Explain the observed waveform.

LIGHT METERS AND PHOTOMETRY

Experiment 41

Name _____

Date _____

Period _____

Lab Partner_____

Purpose:

The purpose of this lab is to measure the intensity of several sources of light, compare them to a standard and determine the efficiency of several sources of light. Also, verify the inverse square law.

Equipment:

Light meter
Meter stick
Various light sources

Procedure:

1. In a completely darkened room, set up each source of light, one at a time.

2. Read the intensity of the source as shown on the light meter, at distances of 0.50 m, 1.00 m, 2.00 m, and 4.00 m from the source.

3. Complete the following table. Find the efficiency of each source at the 2.00 m distance by dividing the source wattage by the meter reading.

sources Intensity of	0.50 m	1.00 m	2.00 m	4.00 m	W/cd Efficiency
25 W					
40 W					
75 W					
100 w					
40 W fluorescent					

4. From the intensity measurements at various distances, verify the inverse square law by completing the table below.

Distances	25w	40w	40w	75w	100w
0.50m					
1.0m					
2.0m					
4.0m					

Questions:

1. What is the effect of doubling the distance from the source?

2. What is the effect of tripling the distance from the source?

3. Do your data show that in fact the intensity varies inversely with the square of the distance from the source?

4. How does the efficiency of the various sources change as the wattage rate increases?

PLANE MIRRORS
Experiment
42

Name _____

Date _____

Period _____

Lab Partner _____

Purpose:

The purpose of this lab is to show how images are formed by plane mirrors.

Equipment:

Plane mirror taped to block of wood
Cardboard
Straight pins
Paper
Ruler and pencil
Protractor

Procedure:

1. Insert three pins through the paper into the cardboard to form a triangle about 6 cm on a side as shown below. Draw lines connecting the three pins.

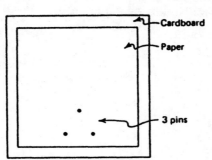

2. Place the mirror about three cm from one pin, positioned so all three pins can be seen from table level in the mirror. Draw a line indicating the position of the mirror.

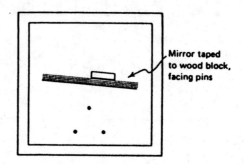

Mirror taped
to wood block,
facing pins

3. Taking one pin at a time, locate the image of the pin by sighting at table level from two different places. Draw a sight line from your eye toward the image from each position as shown. You may use different colored pencils to avoid confusion of many drawn lines.

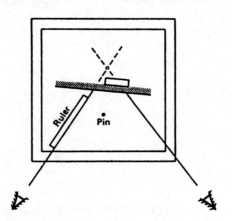

Ruler

Pin

Consider each sight line separately.

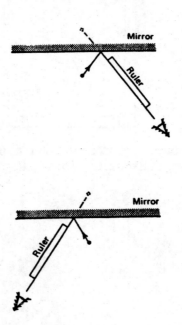

Mirror

Ruler

Mirror

Ruler

4. Lift the mirror and extend the sight lines. They will intersect and determine the location of the virtual image behind the mirror.

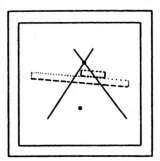

5. Following the same procedure, locate the remaining two pin images. Replace the mirror in exactly the same position as step 2.

6. Connect the three pin images you have located.

7. Measure each of the distances from each pin to the mirror and from each image location to the mirror, and record your measurements in the table below.

8. Construct a perpendicular to the mirror at each point that the sight lines intersect the mirror line for one pin.

9. Measure the angle of incidence and the angle of reflection for each sight line.

Pin Number	Pin Distance From Mirror	Image Distance From Mirror	Angle i	Angle r
1				
2				
3				

Questions:

1. How does the distance of the object from the mirror compare to the distance of the image from the mirror?

2. How do the angles of incidence compare to the angles of reflection?

3. Distinguish the terms: "Real Image" and "Virtual Image".

4. What kind of image do the plane mirrors form?

CONCAVE MIRRORS

Experiment **43**

Name _____

Date _____

Period _____

Lab Partner _____

Purpose:

The purpose of this lab is to determine the focal length of a concave mirror, to verify the mirror formula

$$\frac{1}{f} = \frac{1}{S_o} + \frac{1}{S_i} \quad \text{and} \quad \frac{h_i}{h_o} = \frac{S_i}{S_o}$$

where f = focal length of mirror

S_o = distance of object from mirror

S_i = distance of image from mirror

h_o = height of object

h_i = height of image

and to study the kinds of images formed by concave mirrors.

Equipment:

Concave spherical mirror—focal length approximately 25 cm
2 metersticks
White cardboard image screen
5-15 watt frosted light bulb in socket
Black felt marker

Procedure:

<u>Part 1</u>:

1. Close all window shades except one in the laboratory and project the image of some distant object on the white screen. Adjust the distance from the mirror to the screen to obtain the sharpest image possible. Measure the distance from the mirror to the image on the cardboard screen. This is the focal length of the mirror. Record it in the data table.

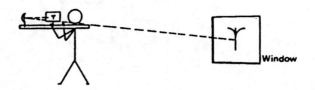

2. Set up the mirror on the meter sticks as shown. Using a black felt marker, draw an object on the bulb before it is turned on (an arrow works well). This is the object. Mount it on one meter stick. Record its distance from the mirror as the object distance, s_o in the table.

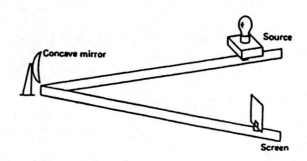

3. Turn on the light and adjust the mirror and slide the screen until a sharp image of the object drawn on the bulb is formed. Record the distance of the focused image from the mirror in the table. This is the image distance s_i. Also measure the height of the object, h_o, and the height of the image, h_i, and record these in the table.

4. Using the same procedure, change the location of the object so the object distance is greater than, equal to, and less than the focal length of the mirror. Measure s_o, s_i and h_i in each case and record in the table.

Trial	f	s_o	s_i	$\dfrac{1}{s_o}$	$\dfrac{1}{s_i}$	$\dfrac{1}{s_o}+\dfrac{1}{s_i}$	$\dfrac{1}{f}$	h_o	h_i	$\dfrac{h_i}{h_o}$	$\dfrac{s_i}{s_o}$
1											
2											
3											
4											
5											

Questions:

1. Is an image formed in each case? If so, describe the image: Virtual or real, Enlarged or diminished, erect or inverted?

2. In which case or cases, if any, is no image formed?

Part 2:

Complete the following diagrams by locating and constructing the image formed.

Use the following principles:

1. A ray through the center of curvature is reflected along itself.

2. A ray parallel to the principal axis is reflected through the focus.

Case 1: Parallel rays

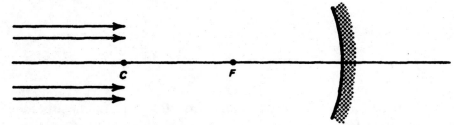

Case 2: Object beyond center of curvature

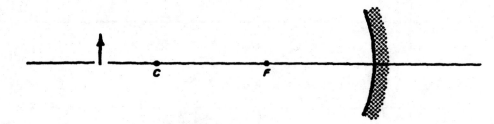

Case 3: Object at the center of curvature

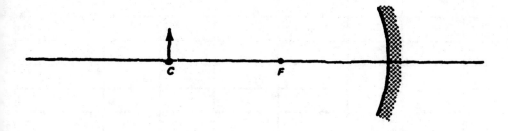

Case 4: Object between focus and center of curvature

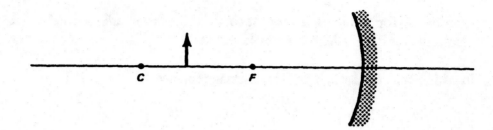

Case 5: Object at focus

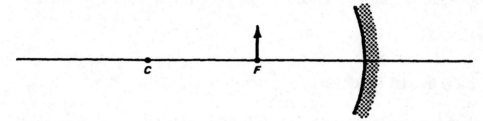

Case 6: Object between focus and mirror

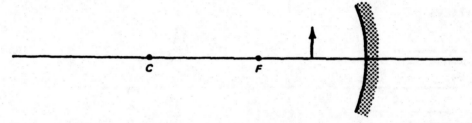

Apply the principles used for concave mirror image constructions to construct the image for the following convex mirror.

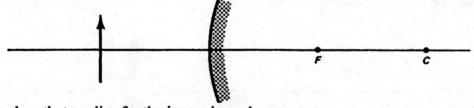

Check the box that applies for the image in each case.

Case	Real	Virtual	Erect	Inverted	Smaller	Same size	Larger	No image
# 1								
# 2								
# 3								
# 4								
# 5								
# 6								
Convex								

INDEX
OF
REFRACTION

Experiment
44

Name _____

Date _____

Period _____

Lab Partner_____

Purpose:

The purpose of this lab is to determine the index of refraction of glass by refracting rays of light.

Equipment:

      Thick glass plate
      Cardboard
      Straight pins
      Ruler
      Paper
      Compass
      Protractor

Procedure:

Using Snell's law

$$n = \frac{\sin i}{\sin r}$$

where n = index of refraction
               i = angle of incidence
               r = angle of refraction

We will locate and diagram the incident and refracted rays; measure the angles; and, with our obtained data, compute the index of refraction.

1. Place the glass plate on the paper and put both on the cardboard with pin #1 inserted next to the glass near a corner of the glass. Outline the glass with a pencil.

2. Near the opposite corner of the glass place pin #2 next to the glass as indicated.

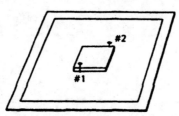

3. Sight through the glass plate at table level from the side of pin #1 and insert pin #3 in line with pin #1 and the image of pin #2 as seen through the glass plate. (Insert pin #3 at least 5 cm away from the plate.)

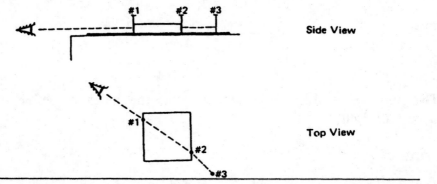

4. Lift the glass plate and with your pencil draw lines connecting pins #1, #2 and #3.

5. Identify the line through pins #1 and #2 as the refracted ray and the line through pins #2 and #3 as the incident ray.

6. Construct perpendiculars to the glass surface line at pin #2.

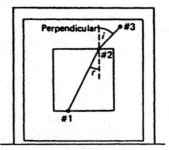

7. Measure the angles of incidence and refraction.

8. Using Snell's law, compute the index of refraction of the glass plate.

9. Make at least three trials and enter your data in the table.

152

Questions:

1. Where the ray passes from air to glass, is the angle of incidence or the angle of refraction larger?

2. Where the ray passes from glass to air, is the angle of incidence or the angle of refraction Larger?

3. If you were to attempt to shoot a fish in water from some point other than directly above, would you have to aim above, below, or directly at the fish?

4. List four everyday examples of the refraction of light that you encounter.

5. Define "critical angle".

6. Calculate the critical angle for water if n = 1.33.

CONVERGING LENSES

Experiment 45

Name _____

Date _____

Period _____

Lab Partner _____

Purpose:

The purpose of this lab is to determine the focal length of a convex lens, to verify the lens formulas

$$\frac{1}{f} = \frac{1}{s_o} + \frac{1}{s_i} \quad \text{and} \quad \frac{h_i}{h_o} = \frac{s_i}{s_o},$$

where

h_o = height of object

h_i = height of image

s_o = distance of object from lens

s_i = distance of image from lens

f = focal length of the lens

and to study the kinds of images formed by convex lenses.

Equipment:

Double convex lens, focal length 10-15 cm
White cardboard screen
Meterstick and supports
5-10 watt frosted light bulb in socket
Black felt marker

Procedure:

<u>Part 1</u>:

1. Close all window shades but one in the laboratory and project the image of some distant object through the lens onto the white screen. Adjust the distance from the lens to the screen to obtain the sharpest image possible. Measure this distance and record it as the focal length of the lens in the table.

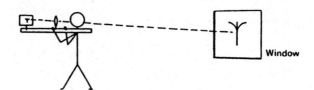

2. Set up the lens and screen on the meterstick as shown. Using a black felt marker draw an object on the bulb before it is turned on (an arrow works well). This is the object. Mount it also on the meterstick. Record the distance from the bulb to the lens in the table as the object distance, s_o.

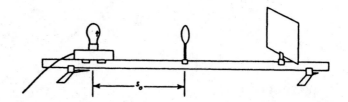

3. Turn on the light and adjust the position of the screen until a sharp image of the object drawn on the bulb is formed. Record the distance of the focused image from the lens, s_i, in the table. Also measure the height of the object, h_o, and the height of the image, h_i, and record these in the table.

4. Using the same procedure, change the location of the object so the object distance is greater than, equal to, and less than the focal length of the lens. Measure s_o, s_i, and h_i in each case and record in the table.

Trial	f	s_o	s_i	$\dfrac{1}{s_o}$	$\dfrac{1}{s_i}$	$\dfrac{1}{s_o}+\dfrac{1}{s_i}$	$\dfrac{1}{f}$	h_o	h_i	$\dfrac{h_i}{h_o}$	$\dfrac{s_i}{s_o}$
1											
2											
3											
4											
5											
6											

156

Questions:

1. Is an image formed in each case?

2. In which case or cases, if any, is no image formed?

3. What kinds of images are formed by converging lenses?

Part 2:

Complete the following diagrams by locating and constructing the image formed.
Use the following principles:

1. A ray through the optical center of the lens is nearly straight and not refracted.

2. A ray parallel to the principal axis is refracted through the focus.

Case 1: Parallel rays

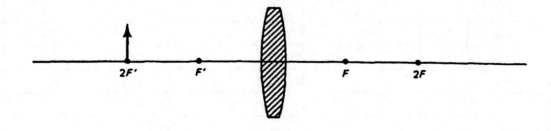

Case 2: Object beyond center of curvature or 2F'

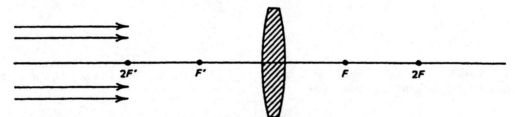

Case 3: Object at 2F'

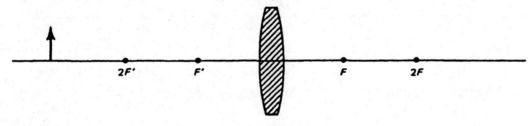

Case 4: Object between F' and 2F'

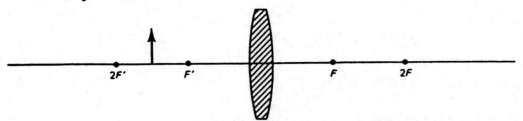

Case 5: Object at focus

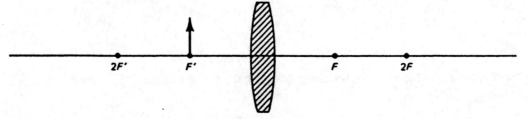

Case 6: Object between F' and lens

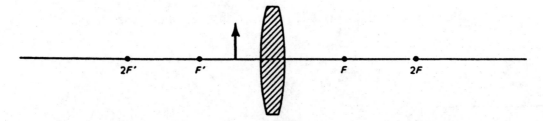

Apply the principles used for convex lens image constructions to construct the image for the following diverging concave lens.

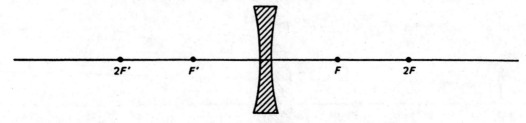

Check the box that correctly describes the image in each case.

Case	Real	Virtual	Erect	Inverted	Smaller	Same size	Larger	No image
# 1								
# 2								
# 3								
# 4								
# 5								
# 6								
Concave								

RADIOACTIVITY

Experiment

46

Name _____

Date _____

Period _____

Lab Partner _____

Purpose:

The purpose of this experiment is to verify the inverse square law of radiation.

Equipment:

    Geiger counter
    Source of beta-gamma radiation
    Calibrated mounting board or ruler

Procedure:

  1. Make sure the Geiger counter is turned off.

  2. Place the source of radiation at a point on the table.

  3. Place the probe of the Geiger tube at a distance of 10 cm from the source.

  4. Note the counts per minute recorded by the instrument.

  5. Move the Geiger counter 5 cm more away from the source.

  6. Note the reading of the instrument.

  7. Repeat the experiment for various distances and record observations.

  8. Measure the background radiation. Move the source from the vicinity of the counter. Note the reading of the Geiger counter. Repeat this 3 or 4 times and get the average background counts.

  9. Use three significant figures for your calculations.

10. Plot the distance versus intensity curve.

11. Plot the distance versus theoretical values of intensity curve.

Type of Geiger counter used   =

Operating voltage   =

Average background count   =

Distance in cms, d	(Intensity) counts per minute, n	Square of distance, d^2	Reciprocal of square of distance, $1/d^2$	Theoretical value of intensity

Questions:

1. Name the three types of natural radioactivity.

2. Would you expect your observations to yield very accurate results? Discuss.

3. A source of gamma radiation produces 20,000 counts per minute on a Geiger counter placed 5 cm from the source. At what distance will the source produce 200 counts/minute? What will be the count when the source is at 15 cm from the tube?

4. Radon has a half-life of 3.82 days. What will be the activity of radon after 15.28 days?

HALF-LIFE OF A RADIOACTIVE ISOTOPE

Experiment 47

Name _____

Date _____

Period _____

Lab Partner _____

Purpose:

The purpose of this lab is to determine the half-life of the given radioactive isotope and use the set of observations to calculate the decay constant.

Equipment:

Geiger counter
Source of radiation (Co-60)
Holder (In-59)
Stop clock

Procedure:

1. Make sure the Geiger counter is off.

2. Place the probe with the Geiger tube on a table.

3. Place the source of radiation on the table about 20 cm from tube. Move the source slowly until the meter reads about one half full scale. Record the exact position of the source. Keep the source in the same place throughout the experiment.

4. Record the reading of counts every 5 minutes.

5. Note background reading without the source and record the net count.

Trials	Time	Back counts/min	Total counts/min	Net counts/min
1	5 min.			
2	10 min.			
3	15			
4	20			
5	25			
6	30			
7	35			
8	40			
9	45			
10	50			
11	55			
12	60			
13	65			
14	70			
15	75			

6. Plot the graph of counts/min versus time on a log-log graph.

7. Calculate the slope and intercept. The slope gives X.

8. Plot N versus time on a linear graph. Draw a smooth curve.

Questions:

1. A source containing 10 microcuries of radioactive Co^{60} is stored away. What percent of the cobalt will disintegrate in one year? The half-life of cobalt 60 is 5.27 yrs.

2. A solution containing 10 microcuries of radioactive P^{32} is obtained on the first of the month. What fraction of radioactive phosphorus remains after 10 days? The half-life of radioactive phosphorus is 14.3 days.

3. A bone emits 4 beta-rays per minute per grain of carbon. How old is this bone? The half-life of carbon is 5,270 years.

4. What is the significance of half-life in treating disease?

DIODES

Experiment

48

Name _____

Date _____

Period _____

Lab Partner _____

Purpose:

The purpose of this experiment is to study the properties of the diode and then use it to convert the alternating current to direct current.

Equipment:

Oscilloscope
A.C. supply
Full-wave rectifier
Diode
Three capacitors
95,000 ohm-resistor

Procedure:

1. Connect the circuit as shown in **Figure 1**. Place the probe of the oscilloscope across the resistor.

2. Record the oscilloscope display on **Graph 1**.

3. Insert the diode in series with resistor as shown in **Figure 2**.

4. Record the oscilloscope display on **Graph 2**.

5. Reverse the direction of the diode in the circuit.

6. Record the oscilloscope display on **Graph 3**.

7. Choose a circuit suitable for an upright display.

8. Place each of the three capacitors, one at a time, across the resistor.

9. Draw the circuits.

10. Record the oscilloscope display in each case on **Graphs 4, 5, and 6**.

11. Connect the full-wave rectifier as shown in **Figure 3**.

12. Record the oscilloscope display on **Graph 7**.

Conclusions:

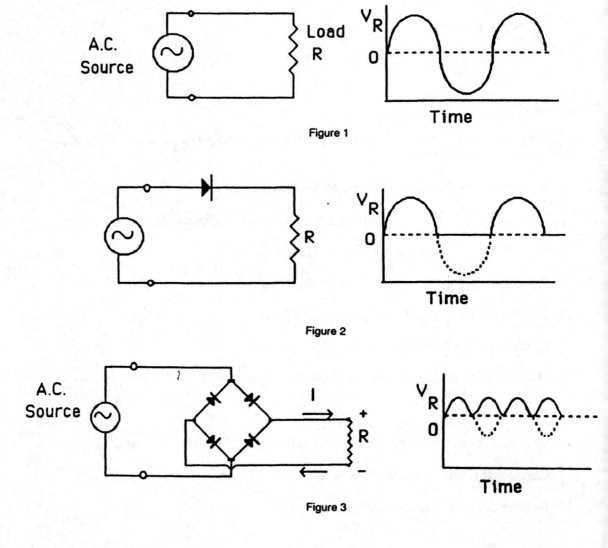

Figure 1

Figure 2

Figure 3

164

Graph 1

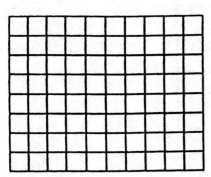

Graph 2

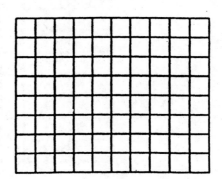

Graph 3

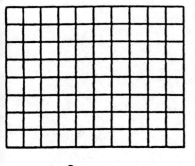

C = _____

RC = _____

Graph 4

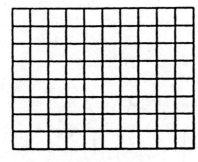

C = _____

RC = _____

Graph 5

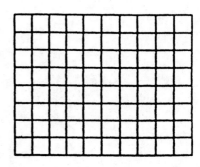

C = _____

RC = _____

Graph 6

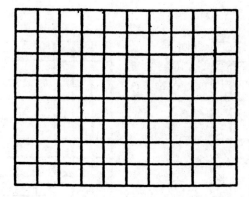

Graph 7

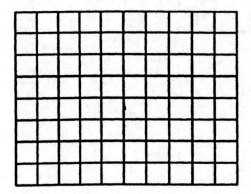

DIODE CHARACTERISTICS

Experiment 49

Name _____

Date _____

Period _____

Lab Partner _____

Purpose:

The purpose of this experiment is to study the properties of semiconductor diodes.

Equipment:

    D.C. power supply 0-20 V
    A.C. power source 12.6 V
    VTVM
    D.C. ammeter 0-10 mA
    CR_1 silicon diode
    R_1 270 ohms, 2 W
    R_2 1K, 1/2 W

Procedure:

1. Set the ohmmeter on the R x 100 scale. Connect the ohmmeter across the diode in the forward bias condition.

2. Measure the forward diode resistance and record it.

3. Connect the ohmmeter across the diode in the reverse bias condition.

4. Measure the reverse diode resistance and record.

5. Explain the difference between these two resistances.

6. Make the connections as shown in **Figure 1.**

7. Starting with 0, increase the voltage until the milliammeter reads 1 mA.

8. Increase the voltage such that the milliammeter reads currents up to 10 mA in steps of 1 mA. Record the values in the table.

9. Draw a graph of voltage versus current and label it.

10. Turn the voltage to 0. Reverse the connections in the diode. Increase the voltage to 20 V. Note the current. Then reduce the voltage to 0.

11. Connect the circuit as shown in **Figure 2.**

12. Measure the current and the voltage with the 270 ohm resistor in the circuit. Record values in the table.

13. Turn the voltage to 0. Replace the 270 ohm resistor with the 1K resistor and repeat step 12. Record the results in the table.

14. Plot the graphs of current versus voltage for steps 12 and 13.

15. Compare the curves. Which is more linear? Can you draw any conclusions from the graphs when the resistance is increased from 270 ohms to 1K?

Questions:

1. How are the P and N type materials made?

2. Name some advantages of semiconductor devices over conventional vacuum tubes.

3. Explain the terms "forward bias" and "reverse bias."

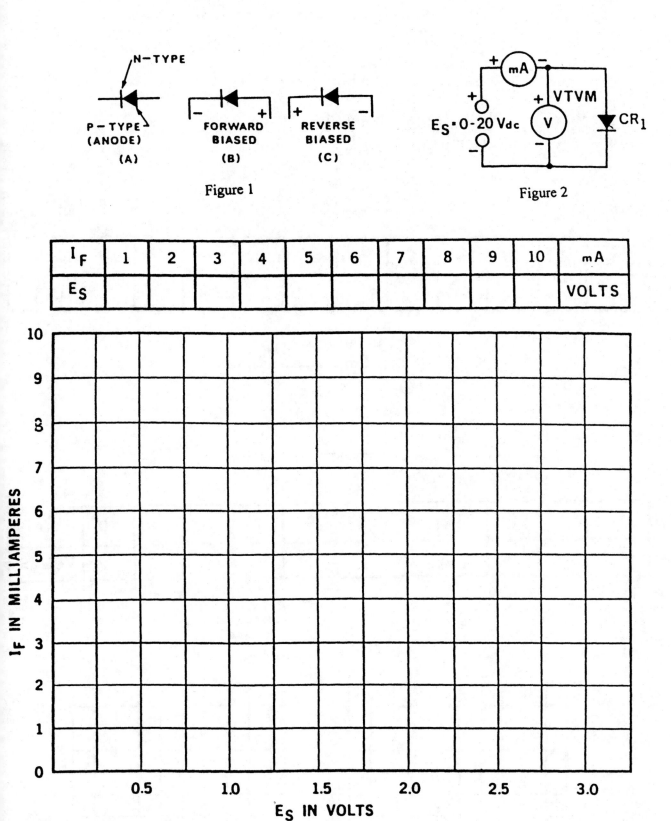

N-TYPE

P-TYPE
(ANODE)
(A)

FORWARD
BIASED
(B)

REVERSE
BIASED
(C)

Figure 1

$E_S \cdot 0\text{-}20\ V_{dc}$

VTVM

CR_1

Figure 2

I_F	1	2	3	4	5	6	7	8	9	10	mA
E_S											VOLTS

I_F IN MILLIAMPERES

E_S IN VOLTS

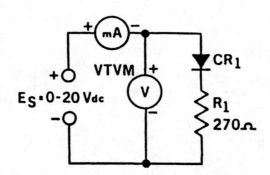

I_F	1	2	3	4	5	6	7	8	9	10	mA
E_S											VOLTS

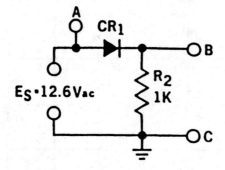

I_F	1	2	3	4	5	6	7	8	9	10	mA
E_S											VOLTS

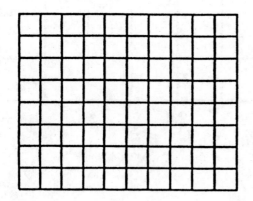
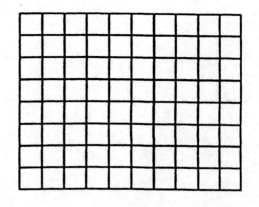

TRANSISTORS Experiment 50

Name _____

Date _____

Period _____

Lab Partner _____

Purpose:

The purpose of this experiment is to identify whether a given transistor is NPN or PNP and whether the transistor is silicon or germanium.

Equipment:

Ohmmeter
PNP transistor
NPN transistor
10 K resistor

Procedure:

1. Use the R x 1K scale on the ohmmeter.

2. Determine which lead of the ohmmeter is positive.

3. Connect the positive lead to the base of the transistor.

4. Measure the resistance.

5. Reverse the polarity.

6. Measure the resistance again.

7. Make a base drawing of each, and label it as PNP or NPN.

8. Connect the resistor and transistor.

9. Measure the voltage from base to emitter.

10. Record the value in volts.

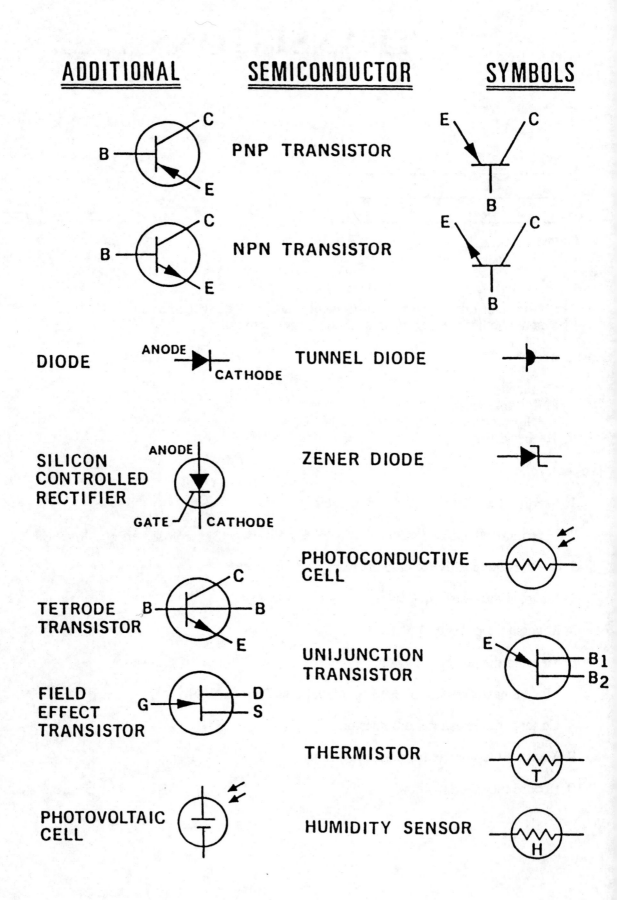

ADDITIONAL   SEMICONDUCTOR   SYMBOLS

PNP TRANSISTOR

NPN TRANSISTOR

DIODE   ANODE   CATHODE   **TUNNEL DIODE**

SILICON
CONTROLLED
RECTIFIER   ANODE   GATE   CATHODE   **ZENER DIODE**

PHOTOCONDUCTIVE CELL

TETRODE
TRANSISTOR   B   C   B   E

UNIJUNCTION
TRANSISTOR   E   B₁   B₂

FIELD
EFFECT
TRANSISTOR   G   D   S

THERMISTOR

PHOTOVOLTAIC
CELL

HUMIDITY SENSOR

SIMPLE HARMONIC MOTION

Name _____

Date _____

Period _____

Lab Partner _____

Purpose:

To study the oscillatory motion of an object suspended from a spring by examining the mathematical relationship between the period and the mass of the object on the spring.

Equipment:

> Spring and support
> Timer
> Weights
> Weight hanger with pointer

Procedure:

1. Find the mass of the spring and the mass of the weight hanger using a beam balance.

2. Place sufficient mass on the weight hanger so the total mass (the mass of the weight hanger plus 1/3 mass of the spring plus the added mass) is approximately 50 gm or .05 kg. Displace the weight hanger slightly in a vertical direction and release. Carefully measure the time required for 20 complete vibrations. Compute and record the experimental value for the period.

3. Repeat step 2 with total mass of 100, 150, 200, 250 and 300 grams.

4. Graph the period vs. the total mass (standard masses and mass of weight hanger plus 1/3 mass of spring) on 2 cycle x 2 cycle log-log paper. (You should get a straight line). The scale on log-log paper is non-linear in the function plotted, but it is linear for the log of the function. The log values for the data are not needed, since the paper itself corrects to a proper linearity. Determine the slope by taking

two points (m_1, T_1) and (m_2, T_2) on the graph (not data points) and solving:

$$S = \frac{\ln T_2 - \ln T_1}{\ln m_2 - \ln m_1} = \frac{\ln (T_2/T_1)}{\ln (m_2/m_1)}$$

5. Find the slope S. Find the percent of error between S and the "correct" value of ½.

6. The y-intercept at $m = 1$ gives the value of $2\pi/\sqrt{k}$. Calculate k.

Questions:

Explain the effect of the following on the period of oscillation:
(a) a stiffer spring
(b) a longer spring
(c) a larger mass
(d) a stronger gravitational field
(e) a larger magnitude of oscillation
(f) air resistance

ELASTICITY OF BONES-- BONE COMPRESSION

Name _____

Date _____

Period _____

Lab Partner _____

Purpose:

The purpose of this experiment is to establish if a relationship exists between density and the compressive forces required to produce a distortion.

Equipment:

Hydraulic Press
Solid steel cylinder
Hollow steel cylinder
Hard wood
Soft wood
Dry bone
Fresh bone

Procedure:

With the exception of steel, run the compression tests before the density tests.

1. Measure the length of the specimen. This may be accomplished by placing the specimen in the hydraulic press and firmly securing the specimen. Then measure the distance between the plates holding the specimen. A fine calibrated ruler will have to be used. Record the length.

2. Apply compressive force by means of the hydraulic press. Record this force.

3. Convert the compressive forces recorded from lbs/in^2 to N/m^2.

4. With the hydraulic press exerting the above force, determine the new length of the specimen.

5. Determine the density for each specimen.

$$\text{Density (g/cm}^3) = \frac{\text{mass}}{\text{Volume}}$$

Determine Young's Modulus (N/m^2) for each specimen.

$$Y = \frac{F/A}{\Delta L/L}$$

	solid steel cylinder	hollow steel cylinder	soft wood	hard wood	dry bone I	dry bone II	fresh bone I	fresh bone II
Original length								
Length after compression								
ΔL								
Compressive force (psi)								
Compressive force (N/m^2)								
Young's Modulus (N/m^2)								
Density (g/cm3)								

Questions:

Compare the density and Young's Modulus of the following:
a) Solid steel, hollow steel cylinders.
b) Solid steel, hollow steel cylinders, compact bone (dry)
c) Hard wood, soft wood, dry bone (I & II), fresh bone (I & II)

1. Compare the ΔL of dry bone and fresh bone. Interpret the comparison.

2. How does bone resemble the comparison materials?

3. Summarize your data.

4. A leg bone has a 1.2 m shaft of bone with an average cross-sectional area of 3 cm^2. What is the amount of shortening when all of the body weight of 700 N is supported on this leg? Y for bone = 1.8×10^{10} N/m^2.

THE LAW OF FLOTATION

Name _____

Date _____

Period _____

Lab Partner _____

Purpose:

The purpose of this lab is to verify the law of flotation using lead shots.

Equipment:

    Measuring cylinder
    Water
    Test tube
    Cotton thread
    Lead shots
    Spring balance measuring in Newtons
    Cotton wool

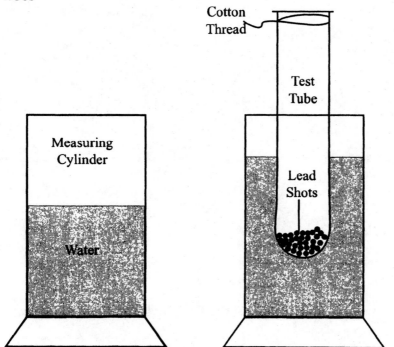

The readings of water level must correspond to the lower meniscus of the level.

Procedure:

The law of flotation states that a floating object displaces its own weight of the fluid in which it floats.

Tie a cotton thread around the neck of a test tube. Fill a measuring cylinder with water to half its capacity. Note the volume of water in the measuring cylinder to start with (V_1). Weigh an empty test tube using the spring balance. Note the weight of the empty test tube (W_0). Put the test tube in water. Add known weight (W_1) of lead shots to the test tube until it floats upright in the water. Note the volume of the water displaced (V_2) in the measuring cylinder. Repeat the experiment five times by adding five different known weights of lead shots.

Data:

Weight of empty test tube = W_o

Trials	Weight of W_0 + Lead Shots (W_1) $W = W_0 + W_1$	Reading of the Measuring Cylinder (cm^3)		Volume of Water Displaced $V_2 - V_1$	Weight of Water Displaced $W_2 = V_2 - V_1/100$
		V_1	V_2		
1					
2					
3					
4					
5					

Questions:

1. Complete the above table.

2. Draw the graph of W vs W_2.

3. Is the graph a straight line?

4. Discuss the relationship of the weight of the floating body to the weight of displaced water in which it floats.

THE TRIANGULAR LAW OF FORCES

Name _____

Date _____

Period _____

Lab Partner _____

Purpose:

The purpose of this lab is to verify the triangular law of forces.

Equipment:

Weights
Three 50 gm hangers
Drawing Board
Sheet of paper
Two pulleys
Thread
Pencil

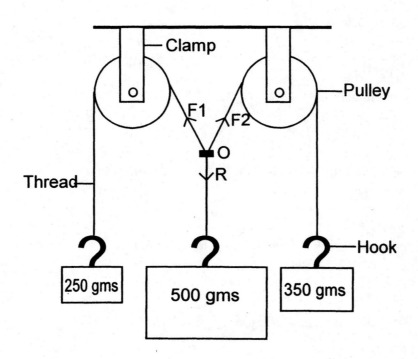

Procedure:

1. Clamp two freely running pulleys to the top of a drawing board.

2. Fix a sheet of paper on the drawing board.

3. Set up the arrangement of drawing board with the sheet of paper attached vertically.

4. Pass a piece of thred over the pulleys.

5. Tie the ends of the thread to 50 gm hangers.

6. Set up the arrangement as shown so that the knot 'O' lies between the two pulleys.

7. Load the hangers with 250 gm, 350 gm, and 500 gm as shown.

8. Mark the positions of the thread as it takes up the position when three forces at 'O' are in equilibrium.

9. Remove the paper from the board and join the crosses to represent forces and their directions.

Data:

Mass m(gms)	250	350	500
Weight W = F(N) F = m/100			

Questions:

1. Complete the table. Repeat the experiment.

2. What is the equilibrant?

3. What is the resultant?

4. Discuss the relationship of equilibrant to the resultant.

5. State the conditions under which the three forces balance.

6. Do these forces drawn form a triangle?

SPECIFIC HEAT
BY ELECTRICAL METHOD

Name _____

Date _____

Period _____

Lab Partner _____

Purpose:

The purpose of this lab is to determine the specific heat of copper by the electrical method.

Equipment:

Electrical supply
Electric immersion heater
Thermometer
Motor oil
Balance
Copper Cylinder with two holes
Stop Watch

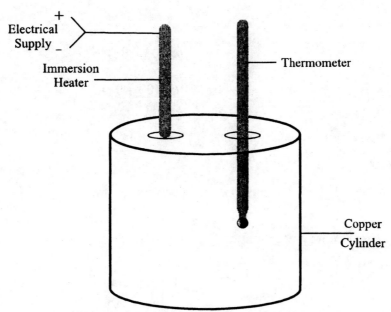

Theory:

When the current is turned on, the electrical energy generated is used to heat up the copper cylinder. Electrical energy is calculated from $\xi=Pt$, where P is the power of the immersion heater and t is the time of passage of current. The heat generated is calculated by using the formula H=$mc\Delta\theta$ where m

is the mass of the copper cylinder, c is the specific heat of copper and $\Delta\theta$ is the rise in temperature. Assuming no heat loss, the electrical energy is equated to the heat generated.

Procedure:

1. Measure the mass of the copper cylinder (m).
2. Insert the immersion heater into one of the holes of the copper cylinder.
3. Fill the copper cylinder with some motor oil through the other hole and insert a thermometer into that hole.
4. Note the initial temperature, θ_1.
5. Connect the heater to the electrical supply.
6. Switch on the current for 5 minutes.
7. Note the power of the heater.
8. Note the time t in seconds.
9. Record the highest temperature reached, θ_2, in t seconds.

Data:

Mass of copper cylinder $\quad m \quad = \quad$ gms.

Power of immersion heater $\quad$ P $\quad = \quad$ watts.

Initial temperature $\quad \theta_1 \quad = \quad$ °C.

Final temperature $\quad \theta_2 \quad = \quad$ °C.

Time of passage of current $\quad t \quad = \quad$ seconds.

$$Pt \quad = \quad mc(\theta_2 - \theta_1).$$

$$c \quad = \quad Pt/m(\theta_2 - \theta_1).$$

Questions:

1. What is the increase in temperature?

2. How much electrical energy was supplied?

3. Knowing the value of specific heat of copper, calculate the heat developed.

4. Can you balance the equation $Pt = mc(\theta_2 - \theta_1)$ in terms of units.

Name _____

Date _____

Period _____

Lab Partner _____

Purpose:

The purpose of this lab is to determine the internal resistance of a cell using a potentiometer.

Equipment:

A battery
A potentiometer
A dry cell (1.5 V)
Two Keys K_1 and K_2
Galvanometer
A Jockey
A resistance box

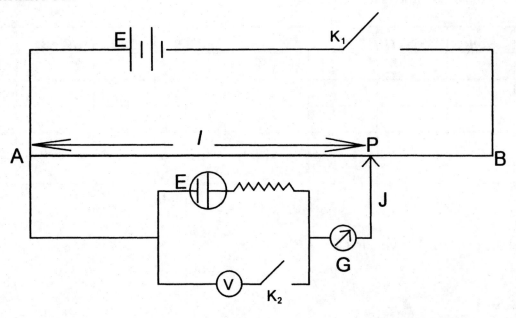

Theory:

The length of the potentiometer wire at the jockey contact is read when the galvanometer reads zero. This is the balance point PO.

At the balance point PO, E is proportional to ℓ_0.

$E = K\ell_0$ where K is a constant. Also at the balance point P, V is proportional to ℓ and

$$V = K\ell \text{ where K is a constant.}$$

If I is the current through the circuit, V = IR. E = I(R+r), where r is the internal resistance of the cell.

$$\frac{V}{E} = \frac{IR}{I(R+r)} = \frac{R}{R+r} = \frac{K\ell}{K\ell_o} = \frac{\ell}{\ell_o}.$$

$$\frac{1}{R} = \frac{I_o}{I}\left(\frac{1}{r}\right) - \frac{1}{r}$$

Procedure:

1. Connect a battery, K_1, and the potentiometer in series as shown.
2. Connect a dry cell (E) of unknown internal resistance r to a resistor and K2 in series.
3. Connect the galvanometer, G, as shown.
4. Close K_1 and open K_2.
5. Find the point Po of the potentiometer.
6. Note the length ℓ_o.
7. Note the EMF of the cell (E).
8. Repeat the measurement of ℓ_o.
9. Take the average value of ℓ_o.
10. Close both K_1 and K_2.
11. Use a resister of 2 ohms, find another balance point P on the wire when the galvanometer reads zero.
12. Note the lengths ℓ when the resistance of the resister is varied to 2Ω, 4Ω, 6Ω, 8Ω, and 10Ω.
13. Record the observations in the table.

Ω	$\ell_o(m)$	Trial 1	Trial 2	Average ℓ_o	R(Ω)	$\ell(m)$	1/R = Ω^{-1}	$\ell_o/\ell = m^{-1}$
2								
4								
6								
8								
10								

Questions:

1. Plot the graph of 1/R versus 1/ℓ.

2. Find the y intercept.

3. What is the meaning of y intercept?

4. Calculate r from the y intercept.

184

WHEATSTONE'S BRIDGE

Name _____

Date _____

Period _____

Lab Partner _____

Purpose:

The purpose of this lab is to determine the resistance of a wire using a Wheatstone bridge.

Equipment:

Galvometer
Resister
Meter bridge
Wire
Jockey key
Standard resister of 2Ω
Nichrome wire 20 cm long
2 V battery

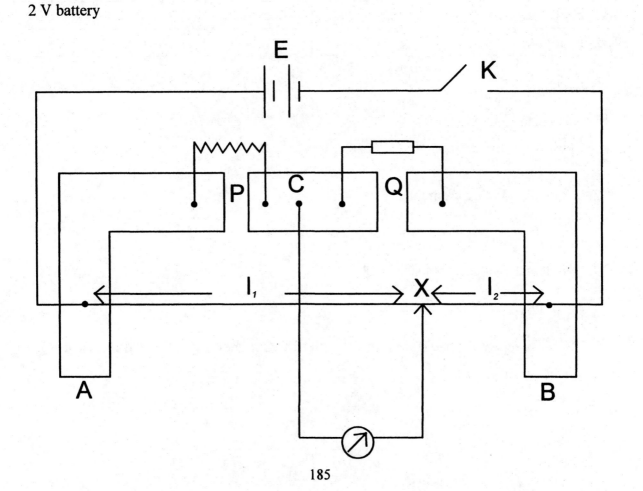

Theory:

Wheatstone bridge is a meter bridge, 100 cms long stretched along a ruler. When the bridge is balanced, i. e., the galvanometer reads zero, therefore:

$$\frac{P}{Q} = \frac{\ell_1}{\ell_2}.$$

P is a known resistance. Q is the unknown resistance we wish to determine. ℓ_1 and ℓ_2 can be measured.

Procedure:

1. Connect a battery to the terminals A and B of the potentiometer wire through a key K.
2. Connect P and Q in the gaps of the Wheatstone bridge.
3. Connect a galvanometer, and a jockey in series with C. C is midway between P and Q.
4. Close the switch.
5. By moving the jockey, find the point X along the wire where galvanometer reads zero.
6. Measure $AX(\ell_1)$ and $BX(\ell_2)$.
7. Open the switch.
8. Interchange P and Q.
9. Repeat the experiment and measure AX and BX again.

Data:

	P (Ω)	(cm)	(cm)	ℓ_1/ℓ_2	$Q = \ell_1/\ell_2(P)$
Experiment 1					
Experiment 2					

Questions:

1. Calculate Q in experiment 1.

2. Calculate Q in experiment 2.

3. What is the value of resistance Q?

You can repeat the same experiment with different unknown values of resistance Q.

FARADAY'S FIRST LAW
OF ELECTROLYSIS

Experiment

58

Name _____

Date _____

Period _____

Lab Partner _____

Purpose:

The purpose of this lab is to verify Faraday's first Law of Electrolysis.

Equipment:

Electrodes (Two similar copper plates)
Copper Sulfate solution
Beaker
Ammeter (0 – 2 A)
Rheostat
Accumulator 6V
Key K
Balance

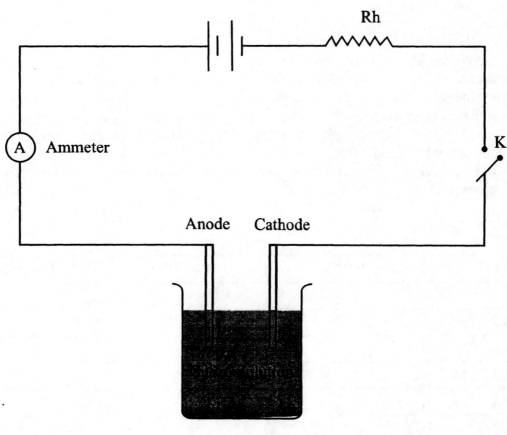

Theory:

Electrolysis is a process of breaking down of elements in a given solution by electric current.

Faraday's 1st Law of Electrolysis states that the mass of an element deposited by electrolysis is directly proportional to the current and time. Therefore, if m is the mass deposited by electrolysis, I is the current and t is the time,

m is directly proportional to It and

therefore, $m = Zit$, where Z is a constant of proportionality. This constant is called "electro-chemical equivalent."

$$Z \quad = \quad \frac{m}{It} \quad (KgA^{-1}s^{-1})$$

But It = quantity of electricity, Q, in coulombs, since I = Q/t. Therefore:

$$Z \quad = \quad \frac{m}{Q} \quad (Kgc^{-1}).$$

This leads us to the definition of electrochemical equivalent (Z). The electrochemical equivalent of a substance is the mass of it liberated or dissolved during electrolysis per unit electric charge that passes into the solution.

Procedure:

1. Use a dummy cathode to adjust a small current of 0.6 A using a rheostat.
2. Open key K.
3. Clean the copper cathode with emery paper, wash it and dry it.
4. Measure the mass m_1 of copper cathode.
5. Replace the dummy cathode with actual cathode.
6. Close the key K.
7. Pass the current for 10 minutes
8. Open key K.
9. Remove the cathode, wash it in distilled water, and dry it by holding it above a small flame.
   Note: Do not hold the cathode near to a flame. The deposited copper may get oxidized.
10. Measure the mass, m_2, of the cathode, after passing current for 10 minutes.
11. Connect the same electrode back into the circuit.
12. Repeat the experiment by increasing current to 0.8 A, 1.0 A, 1.20 A and 1.40 A.
13. Measure the new mass of cathode each time after passing currents for 10 minutes.
14. Keep the current fixed at 1.0 A.
15. Repeat the experiment for various intervals of current and measure the mass of copper electrode for various intervals of time.

Data:

Table 1

I(A)	m_1(gm)	m_2(gm)	$m=(m_2-m_1)$gm
0.6			
0.8			
1.0			
1.2			
1.4			

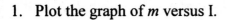

Table 2

t(minutes)	m_1(gm)	m_2(gm)	$m = (m_2 - m_1)$gm
10			
20			
30			
40			

Questions:

1. Plot the graph of m versus I.

2. What is the slope of the graph?

3. How is m related to I?

4. Plot the graph of m versus t.

5. What is the slope of the graph?

6. How is m related to t?

7. How is m related to I and t?

CHARGING AND DISCHARGING A CAPACITOR

Name _____

Date _____

Period _____

Lab Partner _____

Purpose:

The purpose of this lab is (1) to charge a capacitor through a resistor and (2) to discharge a capacitor through a resistor.

Equipment:

16V – 1000 µF electrolytic capacitor
1 KΩ resistor
Two-way switch
Voltmeter 0-5 V
D. C. Source
Watch

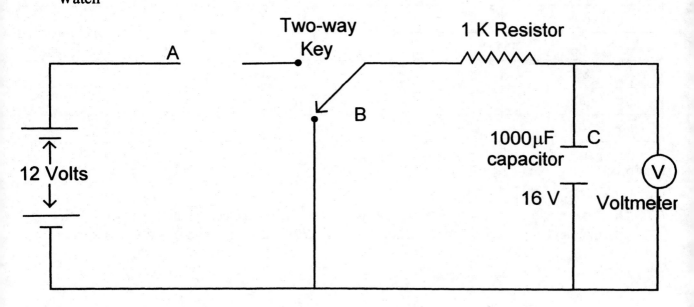

Theory:

A capacitor is a device for storing electric charge. A capacitor charges up to a value of the power source used to charge. Resistor slows down the charging and discharging process.

Make sure the polarities of a capacitor are connected properly. The voltage applied should not be greater than indicated. Otherwise the capacitor will burn up.

Procedure:

1. Connect a 12 V d. c. supply, two-way key, 1 K resistor and 1000 µF capacitor in series as shown. Connect a voltmeter in parallel to the capacitor.
2. Close the key at A. Note the voltmeter reading across the capacitor every 10 seconds using . stop watch until the voltmeter reading is steady.
3. Open the key.
4. Close the key at B. Note the voltmeter reading across the capacitor every 10 seconds until the voltmeter reading is steady.
5. Open the key.

Note: The capacitor is charged to the source voltage and is independent of the resister connected in series to it.

The procedure in (A) will charge the capacitor and in (B) will discharge the capacitor. Record your observations in the table below.

Charging		Discharging	
Time *t* (seconds)	Voltmeter reading (V) Volts	Time *t* (seconds)	Voltmeter Reading V (Volts)
0		120	
10		130	
20		140	
30		150	
40		160	
50		170	
60		180	
70		190	
80		200	
90		210	
100		220	

Questions:

1. Draw the graph of voltmeter reading (V) (potential difference across the capacitor) vs time (*t*) for charging and discharging the capacitor through a resistor. Plot the graph on the same x and y axes.

2. What happens to the p. d. across the capacitor when the switch is closed at A?

3. What happens to the p. d. across the capacitor when the capacitor is fully charged and then disconnected at A?

4. What happens to the p. d. across the capacitor when the switch is closed at B?

5. Calculate the time constant (CR) for charging a capacitor. C is the capacitor and R is the resistor.

6. Discuss the relationship of the maximum p. d. across the capacitor and the voltage of the power source when the switch is closed at A for 100 seconds.

7. What happens to the p. d. across the capacitor ultimately when the switch is closed at B for 100 seconds?

Name _____

Date _____

Period _____

Lab Partner _____

Purpose:

The purpose of this lab is to determine the angle of minimum deviation and refractive index of the material of the prism.

Equipment:

White sheet of paper
Four drawing pins
A glass prism
A pencil
A protractor
Optical board

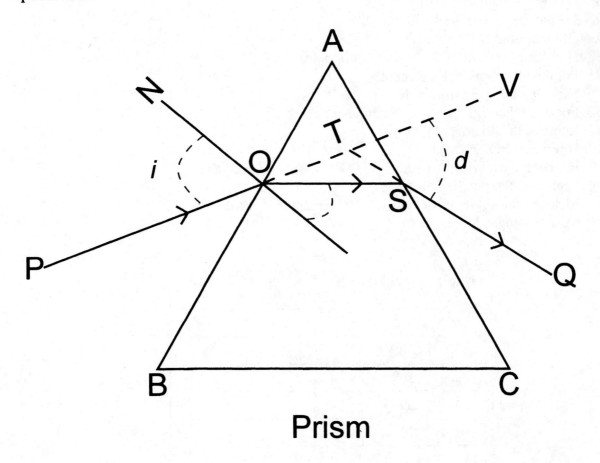

Prism

Theory:

Consider a ray PO incident on face AB at O. It is refracted along SQ after passing through the prism ABC. If the ray PO had propagated without refraction, it would have proceeded towards the point V. Produce QS backwards to meet PV at T. Ray SQ seems to have originated at T. Angle VTQ is the Angle of Deviation (d). By definition, the angle BAC or angle A is the angle of the prism. When the angle of incidence varies between $0°$ and $90°$, there is one angle of incidence at which the angle d is at a minimum. This is called Angle of Minimum Deviation (D) and d (angle of deviation) becomes D.

$$A = 2r$$

$$d = 2\,(i - r)$$

$$A + d = 2r + 2i - 2r = 2i$$

$$\text{Therefore:} \quad i = (A + d)/2$$

$$\text{and } r = A/2.$$

The refractive index $\quad \mu = \dfrac{\text{Sin } i}{\text{Sin } r} = \dfrac{\text{Sin }(A + d)/2}{\text{Sin }(A/2)}.$

When angle of incidence increases, angle of deviation decreases, reaches a minimum and then starts increasing again.

Procedure:

1. Fix the white sheet of paper on the soft board using drawing pins.
2. Place the prism on the sheet and trace the edges of the prism with a sharp pencil.
3. Remove the prism and complete the triangle ABC.
4. Locate the midpoint of AB. It is O.
5. Locate the midpoint of AC. It is S.
6. Draw normal N to AB at O.
7. Draw angles PON, of $30°$ using a protractor.
8. Fix the object pin at P and another at O.
9. View the two pins from side AC.
10. Fix two other pins at S and Q which appear to be along the line of sight PO.
11. Remove the prism and pins.
12. Draw rays PO, OS, SQ.
13. Repeat the experiment with angle PON of $40°$, $50°$, $60°$ and $70°$.
14. Complete the ray diagram.
15. Measure the angles of deviation for each angle of incidence.
16. Measure angle A.

Data:

Angle of incidence (i)	Angle of deviation (d)
$30°$	
$40°$	
$50°$	
$60°$	
$70°$	

Questions:

1. Plot the graph of *i* versus *d*.

2. Identify the angle of minimum deviation *d*.

3. Note the angle of incidence corresponding to the angle of minimum deviation.

4. Determine the angle of refraction *r* at the angle of minimum deviation, *r* = A/2.

5. Evaluate:

$$\mu = \frac{\text{Sin}\,i}{\text{Sin}\,r} = \frac{\text{Sin}\left(\frac{A+d}{2}\right)}{\text{Sin}\left(\frac{A}{2}\right)}$$

Appendix 1

Theory of Errors

Accuracy and precision are two important factors that should be taken into consideration to make meaning of experimental measurements.

The "accuracy" of an observation is the closeness of the experimental value to the accepted value of the quantity under consideration. The "precision" of an experiment is the reproducibility of observations in different trials.

All measurements are affected by errors. This means that measurements are always subject to some uncertainty. There are many types of errors, such as personal, systematic, instrumental, and accidental errors. Personal errors include mistakes in arithmetic calculations, in recording an observation, in reading an observation, or in reading scale division on an instrument. Another kind of personal error is known as "personal bias," trying to fit data to some preconceived idea or being prejudiced in favor of one observation over the other. Instrumental errors are those introduced by slight imperfections or calibrations of the instruments. Systematic errors are characterized by their tendency to be in one direction only, either positive or negative. Accidental errors are deviations beyond the control of the observer. These errors are due to factors in the environment like jarring, noise, fluctuations in temperature and atmospheric pressure, etc,

It is assumed that in experiments, the apparatus is sufficiently well calibrated so that instrumental errors are negligible and the systematic errors, personal errors, and personal bias are eliminated. Under these conditions, all variations in observations are due to accidental errors beyond one's control.

The terms "exact" and "true" values have no place in the language of measurements. It is proper to speak in terms of "best" values or "most probable" values. A best value is determined from a set of measurements. Most commonly the best value is considered to be the average of a set of measurements. The "average" is simply the arithmetic average of a set of observations.

The difference between the best value and an individual measurement in the set is called the "deviation" of the measurement. The deviations determine the inaccuracies of a result and are commonly called "errors." The use of the word "error" is less desirable because it suggests that there is a perfect measurement.

Many times, the value B of some quantity determined in the laboratory is compared with the handbook value A, which is the accepted value of the quantity. The comparison is usually referred to as the percentage error:

$$\text{Percentage Error} = \frac{B-A}{A} \times 100\%$$

199

The same physical quantity is determined by different types of measurements or by repeated observations. In any case, it is meaningful to calculate the percentage difference between two measurements, A_1 and B_1.

$$\text{Percentage difference} = \frac{A_1 - B_1}{\dfrac{A_1 + B_1}{2}} \times 100\%$$

There is no definite value for the allowable percentage of error. In many cases it is reasonable to expect results within 1 %, while in some cases the error may be 5 % or more, depending upon the apparatus used. However, all measurements should be made with the greatest care, to reduce the error as much as possible.

An Illustration of Drawing a Graph

Mass Added to the Spring (Kgm)	F=mg (Newtons)	Stretching of the Spring (m)
0.20	1.96	0.10
0.40	3.92	0.20
0.60	5.88	0.29
0.80	7.84	0.39
1.00	9.80	0.49
1.20	11.76	0.59
1.40	13.72	0.69
1.60	15.68	0.78
1.80	17.64	0.88
2.00	19.60	0.98

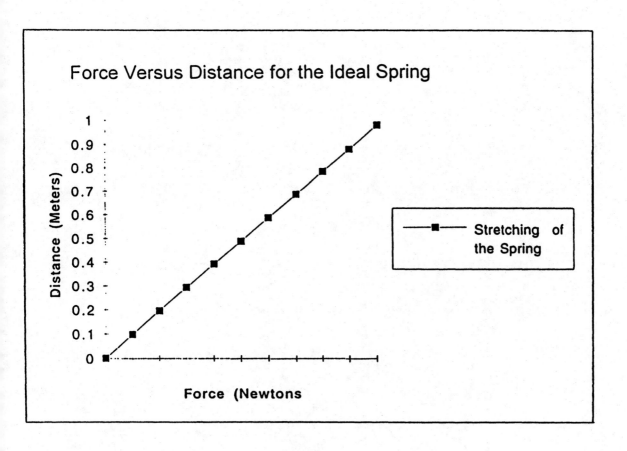

Appendix 2

Graphing Techniques

A mathematical relationship between two measured quantities can often be identified by use of a graph. For example, if a straight line can be drawn through the points, a specific mathematical relationship between the quantities is implied and the relationship is easily obtained.

General Instructions

a) Construct a coordinate by drawing two lines that are perpendicular to each other, crossing at the point called the origin. Both the horizontal and the vertical coordinate axes (the x and y axes) <u>must</u> be labeled with the **name** of the measured quantities, the **symbol**, and the proper **units**, for example, Force, **F**, newtons.

b) Select a suitable scale factor for each coordinate axis so that the resulting graph will appear with a reasonable size and occupy the main part of the graph paper.

c) Plot points and encircle each point with a small circle. To be accurate, the points should be reasonably small in size, but the line that goes through the points may obscure them. Therefore, each point should be encircled.

d) Connect resulting points with a smooth line. The line may be curved or straight. Do not connect points with zigzag line segments.

e) Give a title to the graph. For example, "Force Versus. Distance for the Ideal Spring," or "Power Versus. Current for a Copper Wire." Also place your name and date at the top.

Linear Graph Paper

On linear graph paper, the page is divided evenly in the horizontal and vertical directions. A suitable scale for one of the measured quantities is marked on the horizontal coordinate-axis, and likewise a suitable scale for the other quantity is marked on the vertical coordinate-axis. Each data set is then "graphed" (or "plotted") by finding the point whose coordinates agree with the values of the set.

If the graph of y versus x gives a straight line, the mathematical relationship is $y = mx + b$, where **b**, the value of y when x = 0, is called the y **intercept,** and m is the slope of the line which, can be found by the ratio $(y_1 - y_2)/(x_2 - x_1)$, where (x_1, y_1) and (x_2, y_2) are two arbitrary points <u>on the line</u>, not data points!

Semi-Log and Log-Log Graph Paper

There are two particular cases, both of reasonably common occurrence in physical systems, where the mathematical relationship between measured quantities is readily observed if a log scale is used for the graphing.

a) Many physical phenomena follow the exponential relation: $y = ce^{\alpha x}$, where c and α are constants. This can be written as $\ln y = ce^{\alpha x}$, where c and α are constants. This can be written $\ln y = \ln c + \alpha x$. A plot of $\ln y$ versus x gives a straight line of slope α and intercept of $\ln c$ (at $x = 0$). In the example semi-log plot, **Figure 1**, the slope α is found by the equation $\alpha = (\ln y_2 - \ln y_1) / (x_2 - x_1)$ to be approximately 0.50 and the intercept to be 10. The equation found from the data is $y = ce^{\alpha x}$, which is equal to $10\ e^{0.576x}$.

b) In the same manner, the equation $y = cx^{\alpha}$, where c and α are constants can be written as $\ln y = \ln c + \alpha \ln x$. A plot of $\ln y$ versus $\ln x$ results in a straight line of slope α and intercept $\ln c$ (at $\ln x = 1$). In the example log-log plot, **Figure 2**, the slope α is found using $\alpha = (\ln y_2 - \ln y_1) / (\ln x_2 - \ln x_1)$ to be approximately 0.50 and the intercept to be nearly 10. The equation found from the data is $y = cx^{\alpha} = 10x^{0.5}$.

The log scale is nonlinear and usually broken into "cycles" or repeating sections that do not uniformly increase from 1 to 10. A two-cycle scale, for example, can be numbered from either 1 to 100, or 0.1 to 10, or 10 to 1000, etc. There is no "zero" on a log scale.

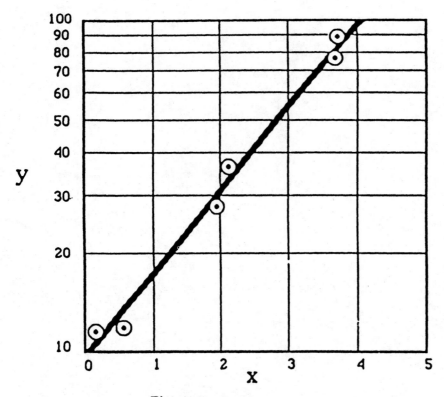

Figure 1   Semi-Log Graph

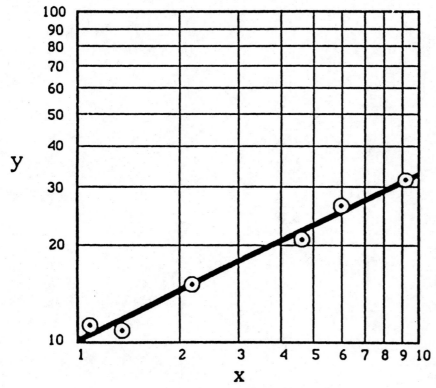

Figure 2   Log-Log Graph

Appendix 3

Radioactivity

The discovery of radioactivity has revolutionized the concept of the nature of the physical world. It has given us the means by which we can study the innermost, central part of the atom, the nucleus.

In 1896, the French physicist Becquerel discovered that minerals containing uranium gave out radiation, penetrated an envelope of black paper, and affected a photographic paper. The rays produced ionization in air and also discharged electrified bodies. This property of uranium was also exhibited by a number of heavy elements. The name "radioactivity" was applied to it. Madam Curie and her husband Pierre Curie discovered polonium and radium, which were much more radioactive than uranium.

The nature of radiation from radioactive substances was discovered by Rutherford and co-workers. The radiation consisted of 3 kinds of rays: alpha, beta, and gamma rays.

Under the influence of a magnetic field, alpha rays were deflected to one side, beta rays to the opposite side, and gamma rays went undeflected. Alpha particles are now identified as the nuclei of helium atoms, consisting of 2 positive charges. The beta particles are the same as electrons. The gamma rays consist of electromagnetic waves of extremely short length.

The radioactive decay is characterized by the equation

$$N = N_0 \, e^{-\lambda t}$$

where N_0 is the number of nuclei present at time $t = 0$, N the number of nuclei after a time t seconds, and λ is the decay constant.

Radioactive decay is represented by "half-life (τ)." This is defined as the time interval in which the nuclei disintegrates to half its initial value.

$$\tau = \lambda/0.693$$

The intensity of the rays emitted from a radioactive substance decreases as the distance from the source increases. This intensity follows the "inverse square law of Radiation." This law states that the intensity of the rays emitted by a radioactive source decreases as the inverse square of the distance from the source.

Appendix 4

Radioisotopes

The nucleus of any atom is made of protons that are positively charged and neutrons that are not charged. The net charge on a nucleus is positive.

Isotopes are a pair of nuclei that have the same number of protons but different numbers of neutrons. For example, Carbon 12 has 6 protons and 6 neutrons. Carbon 13 has 6 protons and 7 neutrons. Carbon 12 and Carbon 13 are isotopes.

Certain combinations of protons and neutrons produce stable isotopes, i.e., isotopes that do not change into other elements. Other combinations produce unstable isotopes, or radioactive isotopes, i.e., isotopes that give off alpha, beta, and gamma rays and change into other elements. Such radioactivity occurring in nature on its own is called natural radioactivity. A radioactive isotope is one that spontaneously disintegrates into a new element by the emission of radiation. Radioisotopes can also be produced artificially by bombarding stable elements with high energy atomic particles.

These radioisotopes find important therapeutic medical applications today in cancer treatment. The whole new branch of nuclear medicine deals with these isotopes.

The concept of half-life also plays a role in archaeology and medicine. The procedure is known as carbon dating. The half-life of carbon is 5270 years. All living tissues emit 16 beta rays per gram of carbon per minute.

Let us say that we find a bone that emits 2 beta rays per minute per gram of carbon. How old is this bone? The answer can easily be obtained from the half-life.

At time $t = 0$, a gram of carbon emits 16 beta rays per minute.

At time $t = 5,270$ years (1 half-life), a gram of carbon will emit 8 beta rays per minute.

At time $t = 10,540$ years (2 half-lives), a gram of carbon will emit 4 beta rays per minute.

At time $t = 15,810$ years (3 half-lives), a gram of carbon will emit 2 beta rays per minute.

Thus the bone is 15,810 years old.

Appendix 5

Cathode Ray Oscilloscope and the Diode

A cathode ray oscilloscope (CRO) is a two-dimensional electronic graphing device. It plots electronically the potential across its sensing probe versus time.

The potential of an alternating current wall outlet varies as a sine wave at frequency 60 Hz.

The semiconductor diode is a circuit element that allows current to flow only in one direction. The diode resistance is either very low or very high depending on the direction the current wants to flow.

The insertion of a diode in series with the resistor will prevent the current from flowing in one direction only, thus converting the alternating current to direct current. The direct current will be very "bumpy," or, pulsating direct current, known as a half wave rectifier circuit.

The insertion of a capacitor across the resistor smooths out bumps, provided the time constant RC is long enough.

The full wave rectifier has 4 diodes.

Appendix 6

Semiconductors and Transistors

A semiconductor is a solid or liquid electronic conductor. Its resistivity is between that of metals and that of insulators. The electronic charge carrier concentration increases with increasing temperature over some temperature range. The resistance has a negative temperature coefficient over most of the practical temperature range.

In a conductor, an electron is removed from one end, leaving a positively charged hole that attracts the adjacent electron, which also leaves a hole. The hole moves through the conductor and is finally filled by an electron from the source at the opposite end.

The new giant of the electronic age is the transistor. The current flow in a transistor may be either electrons or holes, depending on the type of material.

To make a semiconductor, a basic material such as germanium or silicon is "doped" with a minute quantity of impurity. Impurities such as arsenic or antimony are pentavalents. When these are added to germanium, the number of free electrons is increased. Conduction in this type of crystal is by "negative carriers or electrons." It is called N-type crystal.

If trivalent impurities such as aluminium, gallium, or indium are added to germanium, an excess of holes is created. Conduction in this type of crystal is by "holes or positive carriers." It is called P-type crystal.

An interesting phenomenon occurs when N and P type crystals are joined to form a diode. If the positive terminal of a battery is connected to P crystal and the negative terminal to N crystal, the diode is **forward biased**.

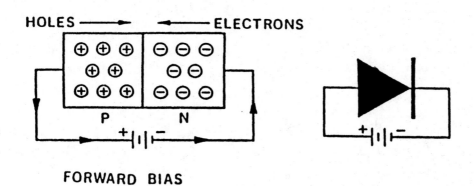

FORWARD BIAS

Conduction in an N crystal is by electrons. They join the holes of P crystal at the junction. Conduction in a P crystal is by holes. The holes are injected into the P crystal as an electron leaves and travels to the positive terminal of the battery. Then the terminal is **reverse biased** and very little current flows in the circuit.

A transistor is an active semiconductor device, usually made of silicon or germanium having three or more electrodes. The three main electrodes used are the emitter, the base, and the collector. The region between an emitter and a collector is called the base.

A transistor with a P-type base and an N-type collector and emitter is called an NPN transistor.

A transistor having two P-type regions separated by an N-type region is called a PNP transistor.

When a small forward bias is applied to the first junction and a large reverse bias to the second junction, the system behaves much like a vacuum tube diode.

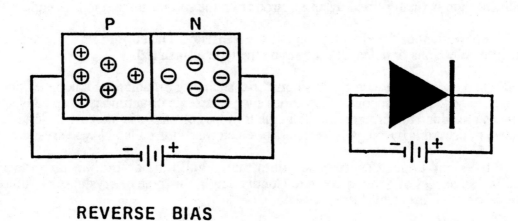

REVERSE BIAS

Appendix 7

Testing of a Transistor

When dealing with an unknown transistor, there are a number of tests that can be performed to give information about the transistor. There are several configurations that are generally recognizable. These are shown in **Figure 1.**

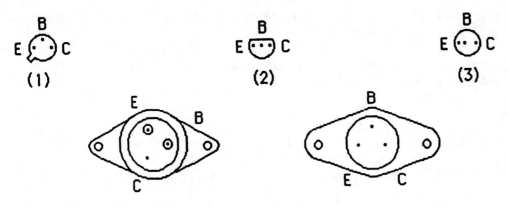

ALL BOTTOM VIEWS

Figure 1

Notice in (4) that the collector is connected to the case. Most of the heat is produced in the collector-base junction since it has the greater voltage across it, and connecting the collector directly to the case helps provide for the dissipation of the heat. In (1) you see the tab nearest the emitter lead. If the transistor does not have a tab, it may have a colored dot on the case near the collector lead. If a transistor does not have any lead obviously connected to the case but has a fourth lead, that lead is probably connected to the case. This is so the case can be grounded to provide isolation (shielding), which is particularly useful in high frequency applications. This lead can be quickly located by an ohmmeter check for zero resistance to the case.

An ohmmeter on the R x 1K scale can be used to determine whether the transistor is an NPN or a PNP transistor. First determine which lead of the ohmmeter is positive. This can be done most easily with another voltmeter. If you believe you have an NPN transistor, connect the positive lead to the base. Then each junction should be forward biased, and the resistance measured should be considerably below the infinity marking. If both measurements show at or near infinity, reverse the polarity. If both are now relatively low, you have a PNP transistor. You should check in both directions to make sure you have a good transistor.

If you find a low resistance in one direction, and infinite resistance in the other direction, either you do not have the base or you have a bad transistor. If you find a near-zero reading, you probably are looking at a "burned through" junction and will probably find a near-zero reading in the reverse direction also. An infinite reading in both directions could mean an "open" junction (no conduction), or you have made connections to emitter and collector junctions. That way you are looking across two junctions, one of which will be in the reverse direction.

Check each of your transistors and lay aside any bad one you find. (Save them for the next student to find!) Make a base drawing of each good transistor, label it with the number or other identifying mark, if any, and label it as PNP or NPN.

Whether the transistor is silicon or germanium can be determined by means of the circuit in **Figure 2**. Measure the voltage from the base to the emitter. If it is a half volt or more, it is a silicon transistor. If it is only 0.3 or 0.4 volts, it is germanium. Germanium transistors are much more temperature sensitive and must be operated at generally lower power levels than silicon transistors.

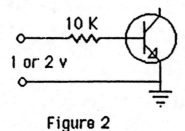

Figure 2

Appendix 8

The Formal Report

The ability to write a scientific report is essential to any scientist. The knowledge gained by a scientist in his/her laboratory is of no use unless it is reported in understandable form and communicated to the scientific world. The format suggested here meets the basic requirements of reports used by many companies and most journals.

One experiment will be chosen each semester for writing a formal report. This report will be considerably longer and more detailed than the general report. The formal report will be due at the time specified by the instructor. The formal laboratory report will enable a student to attain some experience in preparing a manuscript for publication.

Use language that conveys your meaning in the simplest words. Length is not an important factor, the report should be kept as brief as possible, with all essential information included. Be sure that all parts of the report are written in past passive. Write "Such and such was done," rather than "We did such and such."

The formal report should consist of the following sections:

Folder : The lab report should be placed in a plastic or paper folder.

Title Page : This page should contain the same information as contained on the heading of the regular report form and nothing else.

Second Page : The second page must be the first tear-out lab sheet containing the original measurements and observations. For the report to be accepted for grading, this sheet must be stamped or initialed by your instructor prior to your leaving the laboratory: the original data contained on this sheet must be included in an appropriate manner in the body of the report.
Each page thereafter should have at the top the experiment number, title of the experiment, and name of the student. Each section or subsection should be clearly labeled.

Purpose : The objective should be restated in the student's own words.

Apparatus : The apparatus used must be listed.

Procedure : The measurements and calculations that were made should be described in the student's own words.

Data and Results : The data and results should be recorded in tabular form. All graphs and diagrams should be included in this section.

Discussion : A summary of the basic theory involved in the experiment, along with the specific equations used in data analysis, should be given. One sample of each of the types of data substitutions should be carried out; that is, the equations should be given, the values substituted in with their proper units, and the final results shown with simple arithmetic steps omitted. Be sure to include the units.

Questions : Questions should be written out and then answered using sentences.

Conclusion(s): After careful consideration of the experiment, i.e., its purpose, the precision of measurements, the steps in its performance, the results, etc., the student should be able to come to some conclusions. One kind of conclusion might involve a correlation of theory and experiment. A brief summary of the numerical results of the measurements should always be included, even though these results may be clearly evident in the data and results. Any reader who wants to know the results obtained in a research should be able to find them in concise form in the conclusion.

Probable sources of error should be considered here. Any comments about the experiment that do not fit elsewhere may also be made here, i.e., a critique on the equipment used, procedures involved, information supplied, etc.

Appendix 9

TABLES

Table 1   English Weights and Measures

Units of length

Standard unit—inch (in. or ")

12 inches	= 1 foot (ft or ')
3 feet	= 1 yard (yd)
5½ yards or 16½ feet	= 1 rod (rd)
5280 feet	= 1 mile (mi)

Units of weight

Standard unit—pound (lb)

16 ounces (oz)	= 1 pound
2000 pounds	= 1 ton (T)

Units of volume

Liquid

16 ounces (fl oz)	= 1 pint (pt)
2 pints	= 1 quart (qt)
4 quarts	= 1 gallon (gal)

Dry

2 pints (pt)	= 1 quart (qt)
8 quarts	= 1 peck (pk)
4 pecks	= 1 bushel (bu)

Table 2   Conversion Table for Length

	cm	m	km	in.	ft	mile
1 centimeter =	1	10^{-2}	10^{-5}	0.394	3.28×10^{-2}	6.21×10^{-6}
1 meter =	100	1	10^{-3}	39.4	3.28	6.21×10^{-4}
1 kilometer =	10^5	1000	1	3.94×10^4	3280	0.621
1 inch =	2.54	2.54×10^{-2}	2.54×10^{-5}	1	8.33×10^{-2}	1.58×10^{-5}
1 foot =	30.5	0.305	3.05×10^{-4}	12	1	1.89×10^{-4}
1 mile =	1.61×10^5	1610	1.61	6.34×10^4	5280	1

Table 3   Conversion Table for Area

Metric	English
1 m² = 10,000 cm²	1 ft² = 144 in²
= 1,000,000 mm²	1 yd² = 9 ft²
1 cm² = 100 mm²	1 rd² = 30.25 yd²
= 0.0001 m²	1 acre = 160 rd²
1 km² = 1,000,000 m²	= 4840 yd²
	= 43,560 ft²
	1 mi² = 640 acres

	m²	cm²	ft²	in²
1 square meter =	1	10^4	10.8	1550
1 square centimeter =	10^{-4}	1	1.08×10^{-3}	0.155
1 square foot =	9.29×10^{-2}	929	1	144
1 square inch =	6.45×10^{-4}	6.45	6.94×10^{-3}	1

1 circular mil = 5.07×10^{-6} cm² = 7.85×10^{-7} in²

1 hectare = 10,000 m² = 2.47 acres

Table 4 Conversion Table for Volume

	Metric	English
	$1\ m^3 = 10^6\ cm^3$	$1\ ft^3 = 1728\ in^3$
	$1\ cm^3 = 10^{-6}\ m^3$	$1\ yd^3 = 27\ ft^3$
	$= 10^3\ mm^3$	

	m^3	cm^3	L	ft^3	in^3
1 m^3 =	1	10^6	1000	35.3	6.10×10^4
1 cm^3 =	10^{-6}	1	1.00×10^{-3}	3.53×10^{-5}	6.10×10^{-2}
1 litre =	1.00×10^{-3}	1000	1	3.53×10^{-2}	61.0
1 ft^3 =	2.83×10^{-2}	2.83×10^4	28.3	1	1728
1 in^3 =	1.64×10^{-5}	16.4	1.64×10^{-2}	5.79×10^{-4}	1

1 U.S. fluid gallon = 4 U.S. fluid quarts = 8 U.S. pints = 128 U.S. fluid ounces = 231 in^3 = 0.134 ft^3
1 L = 1000 cm^3 = 1.06 qt   1 fl oz = 29.5 cm^3
1 ft^3 = 7.47 gal = 28.3 L

Table 5 Conversion Table for Mass

	g	kg	slug	oz	lb	ton
1 gram =	1	0.001	6.85×10^{-5}	3.53×10^{-2}	2.21×10^{-3}	1.10×10^{-6}
1 kilogram =	1000	1	6.85×10^{-2}	35.3	2.21	1.10×10^{-3}
1 slug =	1.46×10^4	14.6	1	515	32.2	1.61×10^{-2}
1 ounce =	28.4	2.84×10^{-2}	1.94×10^{-3}	1	6.25×10^{-2}	3.13×10^{-5}
1 pound =	454	0.454	3.11×10^{-2}	16	1	5.00×10^{-4}
1 ton =	9.07×10^5	907	62.2	3.2×10^4	2000	1

1 metric ton = 1000 kg = 2205 lb

Table 6 Conversion Table for Density

	$slug/ft^3$	kg/m^3	g/cm^3	lb/ft^3	lb/in^3
1 slug per ft^3 =	1	515	0.515	32.2	1.86×10^{-2}
1 kilogram per m^3 =	1.94×10^{-3}	1	0.001	6.24×10^{-2}	3.61×10^{-5}
1 gram per cm^3 =	1.94	1000	1	62.4	3.61×10^{-2}
1 pound per ft^3 =	3.11×10^{-2}	16.0	1.60×10^{-2}	1	5.79×10^{-4}
1 pound per in^3 =	53.7	2.77×10^4	27.7	1728	1

Table 7 Conversion Table for Time

	yr	day	h	min	s
1 year =	1	365	8.77×10^3	5.26×10^5	3.16×10^7
1 day =	2.74×10^{-3}	1	24	1440	8.64×10^4
1 hour =	1.14×10^{-4}	4.17×10^{-2}	1	60	3600
1 minute =	1.90×10^{-6}	6.94×10^{-4}	1.67×10^{-2}	1	60
1 second =	3.17×10^{-8}	1.16×10^{-5}	2.78×10^{-4}	1.67×10^{-2}	1

Table 8 Conversion Table for Speed

	ft/s	km/h	m/s	mi/h	cm/s
1 foot per second =	1	1.10	0.305	0.682	30.5
1 kilometer per hour =	0.911	1	0.278	0.621	27.8
1 meter per second =	3.28	3.60	1	2.24	100
1 mile per hour =	1.47	1.61	0.447	1	44.7
1 centimeter per second =	3.28×10^{-2}	3.60×10^{-2}	0.01	2.24×10^{-2}	1

1 mi/min = 88.0 ft/s = 60.0 mi/h

Table 9 Conversion Table for Force

	N	lb	oz
1 newton =	1	0.225	3.60
1 pound =	4.45	1	16
1 ounce =	0.278	0.0625	1

Table 10 Conversion Table for Power

	Btu/h	ft lb/s	hp	cal/s	kW	W
1 British thermal unit per hour =	1	0.216	3.93×10^{-4}	7.00×10^{-2}	2.93×10^{-4}	0.29
1 foot pound per second =	4.63	1	1.82×10^{-3}	0.324	1.36×10^{-3}	1.36
1 horsepower =	2550	550	1	178	0.746	746
1 kilowatt =	3410	738	1.34	239	1	1000
1 watt =	3.41	0.738	1.34×10^{-3}	0.239	0.001	1

Table 11 Conversion Table for Pressure

	atm	inch of water	mm-Hg	N/m^2 (Pa)	lb/in^2	lb/ft^2
1 atmosphere =	1	407	760	1.01×10^5	14.7	2120
1 inch of water* at 4°C =	2.46×10^{-3}	1	1.87	249	3.61×10^{-2}	5.20
1 millimeter of mercury* at 0°C =	1.32×10^{-3}	0.535	1	133	1.93×10^{-2}	2.79
1 newton per meter² (pascal) =	9.87×10^{-6}	4.02×10^{-3}	7.50×10^{-3}	1	1.45×10^{-4}	2.09×10
1 pound per in² =	6.81×10^{-2}	27.7	51.7	6.90×10^3	1	144
1 pound per ft² =	4.73×10^{-4}	0.192	0.359	47.9	6.94×10^{-3}	1

* Where the acceleration of gravity has the standard value, $9.80 \text{ m/s}^2 = 32.2 \text{ ft/s}^2$.

Table 12 Mass and Weight Density[a]

Substance	Mass density (kg/m³)	Weight density (lb/ft³)
Solids		
Copper	8,890	555
Iron	7,800	490
Lead	11,300	708
Aluminum	2,700	169
Ice	917	57
Wood, white pine	420	26
Concrete	2,300	145
Cork	240	15
Liquids		
Water	1,000	62.4
Seawater	1,025	64.0
Oil	870	54.2
Mercury	13,600	846
Alcohol	790	49.4
Gasoline	680	42.0

	At 0°C and 1 atm pressure	At 32°F and 1 atm pressure
Gases[a]		
Air	1.29	0.081
Carbon dioxide	1.96	0.123
Carbon monoxide	1.25	0.078
Helium	0.178	0.011
Hydrogen	0.0899	0.0056
Oxygen	1.43	0.089
Nitrogen	1.25	0.078
Ammonia	0.760	0.047
Propane	2.02	0.126

[a] The density of a gas is found by pumping the gas into a container, by measuring its volume and mass or weight, and then by using the appropriate density formula.

Table 13 Specific Gravity of Certain Liquids

Liquid	Specific gravity
Benzene	0.9
Ethyl alcohol	0.79
Gasoline	0.68
Kerosene	0.82
Mercury	13.6
Seawater	1.025
Sulfuric acid	1.84
Turpentine	0.87
Water	1.000

Table 14 Conversion Table for Energy, Work, and Heat

	Btu	ft lb	J	cal	kWh
1 British thermal unit =	1	778	1060	252	2.93×10^{-4}
1 foot pound =	1.29×10^{-3}	1	1.36	0.324	3.77×10^{-7}
1 horsepower-hour =	2550	1.98×10^{6}	2.69×10^{6}	6.41×10^{5}	0.746
1 joule =	9.48×10^{-4}	0.738	1	0.239	2.78×10^{-7}
1 calorie =	3.97×10^{-3}	3.09	4.19	1	1.16×10^{-6}
1 kilowatt-hour =	3410	2.66×10^{6}	3.60×10^{6}	8.60×10^{5}	1

Table 15 Heat Constants

	Melting point (°C)	Boiling point (°C)	Specific heat cal/g°C or kcal/kg°C or Btu/lb°F	J/kg°C	Heat of fusion cal/g or kcal/kg	J/kg	Heat of vaporization cal/g or kcal/kg	J/kg
Alcohol, ethyl	−117	78.5	0.58	2400	24.9	1.04×10^{5}	204	8.54×10^{5}
Aluminum	660	2057	0.22	920	76.8	3.21×10^{5}		
Brass	840		0.092	390				
Copper	1083	2330	0.092	390	49.0	2.05×10^{5}		
Glass			0.21	880				
Ice	0		0.51	2100	$8\bar{0}$	3.35×10^{5}		
Iron (steel)	1540	3000	0.115	481	7.89	3.30×10^{4}		
Lead	327	1620	0.031	130	5.86	2.45×10^{4}		
Mercury	−38.9	357	0.033	140	2.82	1.18×10^{4}	65.0	2.72×10^{5}
Silver	961	1950	0.056	230	26.0	1.09×10^{5}		
Steam			0.48	$20\bar{0}0$				
Water (liquid)	0	$10\bar{0}$	1.00	4190			$54\bar{0}$	2.26×10^{6}
Zinc	419	907	0.092	390	23.0	9.63×10^{4}		

Table 16 Coefficient of Linear Expansion

Material	α (metric)	α (English)
Aluminum	$2.3 \times 10^{-5}/C°$	$1.3 \times 10^{-5}/F°$
Brass	$1.9 \times 10^{-5}/C°$	$1.0 \times 10^{-5}/F°$
Concrete	$1.1 \times 10^{-5}/C°$	$6.0 \times 10^{-6}/F°$
Copper	$1.7 \times 10^{-5}/C°$	$9.5 \times 10^{-6}/F°$
Glass	$9.0 \times 10^{-6}/C°$	$5.1 \times 10^{-6}/F°$
Pyrex	$4.0 \times 10^{-6}/C°$	$1.7 \times 10^{-6}/F°$
Steel	$1.3 \times 10^{-5}/C°$	$6.5 \times 10^{-6}/F°$
Zinc	$2.6 \times 10^{-5}/C°$	$1.5 \times 10^{-5}/F°$

Table 17 Coefficient of Volume Expansion

Liquid	β (metric)	β (English)
Acetone	$1.49 \times 10^{-3}/C°$	$8.28 \times 10^{-4}/F°$
Alcohol, ethyl	$1.12 \times 10^{-3}/C°$	$6.62 \times 10^{-4}/F°$
Carbon tetrachloride	$1.24 \times 10^{-3}/C°$	$6.89 \times 10^{-4}/F°$
Mercury	$1.8 \times 10^{-4}/C°$	$1.0 \times 10^{-4}/F°$
Petroleum	$9.6 \times 10^{-4}/C°$	$5.33 \times 10^{-4}/F°$
Turpentine	$9.7 \times 10^{-4}/C°$	$5.39 \times 10^{-4}/F°$
Water	$2.1 \times 10^{-4}/C°$	$1.17 \times 10^{-4}/F°$

Table 18 Conversion Table for Charge

Charge on one electron = 1.60×10^{-19} coulomb

1 coulomb = 6.25×10^{18} electrons of charge

1 ampere-hour = 3600 C

Table 19   Copper Wire Table

Gauge no.	Diameter (mils)	Diameter (mm)	Cross section		Ohms per 1000 ft		Weight per 100 (lb).
			cir mils	in²	25°C (77°F)	65°C (149°F)	
0000	460.0		212.000	0.166	0.0500	0.0577	641.0
000	410.0		168.000	0.132	0.0630	0.0727	508.0
00	365.0		133.000	0.105	0.0795	0.0917	403.0
0	325.0		106.000	0.0829	0.100	0.116	319.0
1	289.0	7.35	83,700	0.0657	0.126	0.146	253.0
2	258.0	6.54	66,400	0.0521	0.159	0.184	201.0
3	229.0	5.83	52,600	0.0413	0.201	0.232	159.0
4	204.0	5.19	41,700	0.0328	0.253	0.292	126.0
5	182.0	4.62	33,100	0.0260	0.319	0.369	100.0
6	162.0	4.12	26,300	0.0206	0.403	0.465	79.5
7	144.0	3.67	20,800	0.0164	0.508	0.586	63.0
8	128.0	3.26	16,500	0.0130	0.641	0.739	50.0
9	114.0	2.91	13,100	0.0103	0.808	0.932	39.6
10	102.0	2.59	10,400	0.00815	1.02	1.18	31.4
11	91.0	2.31	8,230	0.00647	1.28	1.48	24.9
12	81.0	2.05	6,530	0.00513	1.62	1.87	19.8
13	72.0	1.83	5,180	0.00407	2.04	2.36	15.7
14	64.0	1.63	4,110	0.00323	2.58	2.97	12.4
15	57.0	1.45	3,260	0.00256	3.25	3.75	9.86
16	51.0	1.29	2,580	0.00203	4.09	4.73	7.82
17	45.0	1.15	2,050	0.00161	5.16	5.96	6.20
18	40.0	1.02	1,620	0.00128	6.51	7.51	4.92
19	36.0	0.91	1,290	0.00101	8.21	9.48	3.90
20	32.0	0.81	1,020	0.000802	10.4	11.9	3.09
21	28.5	0.72	810	0.000636	13.1	15.1	2.45
22	25.3	0.64	642	0.000505	16.5	19.0	1.94
23	22.6	0.57	509	0.000400	20.8	24.0	1.54
24	20.1	0.51	404	0.000317	26.2	30.2	1.22
25	17.9	0.46	320	0.000252	33.0	38.1	0.970
26	15.9	0.41	254	0.000200	41.6	48.0	0.769
27	14.2	0.36	202	0.000158	52.5	60.6	0.610
28	12.6	0.32	160	0.000126	66.2	76.4	0.484
29	11.3	0.29	127	0.0000995	83.4	96.3	0.384
30	10.0	0.26	101	0.0000789	105	121	0.304
31	8.9	0.23	79.7	0.0000626	133	153	0.241
32	8.0	0.20	63.2	0.0000496	167	193	0.191
33	7.1	0.18	50.1	0.0000394	211	243	0.152
34	6.3	0.16	39.8	0.0000312	266	307	0.120
35	5.6	0.14	31.5	0.0000248	335	387	0.0954
36	5.0	0.13	25.0	0.0000196	423	488	0.0757
37	4.5	0.11	19.8	0.0000156	533	616	0.0600
38	4.0	0.10	15.7	0.0000123	673	776	0.0476
39	3.5	0.09	12.5	0.0000098	848	979	0.0377
40	3.1	0.08	9.9	0.0000078	1070	1230	0.0200

Table 20   Conversion Table for Plane Angles

				rad	rev
1 degree =	1	60	3600	1.75×10^{-2}	2.78×10^{-3}
1 minute =	1.67×10^{-2}	1	60	2.91×10^{-4}	4.63×10^{-5}
1 second =	2.78×10^{-4}	1.67×10^{-2}	1	4.85×10^{-6}	7.72×10^{-7}
1 radian =	57.3	3440	2.06×10^{5}	1	0.159
1 revolution =	360	2.16×10^{4}	1.30×10^{6}	6.28 or 2π	1

Table 21 Trigonometric Ratios

Angle (°)	Sine	Cosine	Tangent	Angle (°)	Sine	Cosine	Tangent
0	0.000	1.000	0.000	45	0.707	0.707	1.000
1	0.017	0.999	0.017	46	0.719	0.695	1.036
2	0.035	0.999	0.035	47	0.731	0.682	1.072
3	0.052	0.999	0.052	48	0.743	0.669	1.111
4	0.070	0.998	0.070	49	0.755	0.656	1.150
5	0.087	0.996	0.087	50	0.766	0.643	1.192
6	0.105	0.995	0.105	51	0.777	0.629	1.235
7	0.122	0.993	0.123	52	0.788	0.616	1.280
8	0.139	0.990	0.141	53	0.799	0.602	1.327
9	0.156	0.988	0.158	54	0.809	0.588	1.376
10	0.174	0.985	0.176	55	0.819	0.574	1.428
11	0.191	0.982	0.194	56	0.829	0.559	1.483
12	0.208	0.978	0.213	57	0.839	0.545	1.540
13	0.225	0.974	0.231	58	0.848	0.530	1.600
14	0.242	0.970	0.249	59	0.857	0.515	1.664
15	0.259	0.966	0.268	60	0.866	0.500	1.732
16	0.276	0.961	0.287	61	0.875	0.485	1.804
17	0.292	0.956	0.306	62	0.883	0.469	1.881
18	0.309	0.951	0.325	63	0.891	0.454	1.963
19	0.326	0.946	0.344	64	0.899	0.438	2.050
20	0.342	0.940	0.364	65	0.906	0.423	2.145
21	0.358	0.934	0.384	66	0.914	0.407	2.246
22	0.375	0.927	0.404	67	0.921	0.391	2.356
23	0.391	0.921	0.424	68	0.927	0.375	2.475
24	0.407	0.914	0.445	69	0.934	0.358	2.605
25	0.423	0.906	0.466	70	0.940	0.342	2.747
26	0.438	0.899	0.488	71	0.946	0.326	2.904
27	0.454	0.891	0.510	72	0.951	0.309	3.078
28	0.469	0.883	0.532	73	0.956	0.292	3.271
29	0.485	0.875	0.554	74	0.961	0.276	3.487
30	0.500	0.866	0.577	75	0.966	0.259	3.732
31	0.515	0.857	0.601	76	0.970	0.242	4.011
32	0.530	0.848	0.625	77	0.974	0.225	4.331
33	0.545	0.839	0.649	78	0.978	0.208	4.705
34	0.559	0.829	0.675	79	0.982	0.191	5.145
35	0.574	0.819	0.700	80	0.985	0.174	5.671
36	0.588	0.809	0.727	81	0.988	0.156	6.314
37	0.602	0.799	0.754	82	0.990	0.139	7.115
38	0.616	0.788	0.781	83	0.993	0.122	8.144
39	0.629	0.777	0.810	84	0.995	0.105	9.514
40	0.643	0.766	0.839	85	0.996	0.087	11.43
41	0.656	0.755	0.869	86	0.998	0.070	14.30
42	0.669	0.743	0.900	87	0.999	0.052	19.08
43	0.682	0.731	0.933	88	0.999	0.035	28.64
44	0.695	0.719	0.966	89	0.999	0.017	57.29
45	0.707	0.707	1.000	90	1.000	0.000	—

Table 22   The Greek Alphabet

Capital	Lowercase	Name
Α	α	alpha
Β	β	beta
Γ	γ	gamma
Δ	δ	delta
Ε	ε	epsilon
Ζ	ζ	zeta
Η	η	eta
Θ	θ	theta
Ι	ι	iota
Κ	κ	kappa
Λ	λ	lambda
Μ	μ	mu
Ν	ν	nu
Ξ	ξ	xi
Ο	ο	omicron
Π	π	pi
Ρ	ρ	rho
Σ	σ	sigma
Τ	τ	tau
Υ	υ	upsilon
Φ	φ	phi
Χ	χ	chi
Ψ	ψ	psi
Ω	ω	omega

Table 23

Strength of Bone and Other Common Materials

Material	Compressive Breaking Stress (N/mm^2)	Tensile Breaking Stress (N/mm^2)	Young's Modulus of Elasticity (X10^{2}N/mm^2)
Hard Steel	552	827	2070
Rubber	0	2.1	0.01
Granite	145	4.8	517
Concrete	21	2.1	165
Oak	59	117	110
Porcelain	552	55	0
Compact bone	170	120	179
Trabecular bone	2.2	0	0.76